SpringerBriefs in Service Science

SpringerBriefs present concise summaries of cutting-edge research and practical applications across a wide spectrum of fields. Featuring compact volumes of 50 to 125 pages, the series covers a range of content from professional to academic.

Typical publications can be:

A timely report of state-of-the art methods

A bridge between new research results, as published in journal articles

A snapshot of a hot or emerging topic

An in-depth case study

A presentation of core concepts that students must understand in order to make independent contributions

SpringerBriefs are characterized by fast, global electronic dissemination, standard publishing contracts, standardized manuscript preparation and formatting guidelines, and expedited production schedules.

The rapidly growing fields of Big Data, AI and Machine Learning, together with emerging analytic theories and technologies, have allowed us to gain comprehensive insights into both social and transactional interactions in service value co-creation processes. The series SpringerBriefs in Service Science is devoted to publications that offer new perspectives on service research by showcasing service transformations across various sectors of the digital economy. The research findings presented will help service organizations address their service challenges in tomorrow's service-oriented economy.

Haiyan Yu

Data Quality Management in the Data Age

Excellence in Data Quality for Enhanced Digital Economic Growth

 Springer

Haiyan Yu
Chongqing University of Posts and Telecommunications
Chongqing, China

ISSN 2731-3743   ISSN 2731-3751   (electronic)
SpringerBriefs in Service Science
ISBN 978-3-031-71870-0   ISBN 978-3-031-71871-7   (eBook)
https://doi.org/10.1007/978-3-031-71871-7

This Springer imprint is published by the registered company Springer Nature Switzerland AG
The registered company address is: Gewerbestrasse 11, 6330 Cham, Switzerland

If disposing of this product, please recycle the paper.

Preface

As a new factor of production, data has become indispensable in fostering the growth of the digital economy. Data markets provide a mechanism for the acquisition of high-quality data (or data goods) and for augmenting data supply. Data market managers and regulatory bodies must establish an environment that incentivizes potential data sellers, particularly those with high-quality data, to participate in the market. Consequently, data scientists and engineers operating within these markets must enhance their proficiency in data quality management.

This book is tailored for data scientists and engineers, as well as for managers of data markets, researchers, and graduate students in quality engineering, statistics, and service science. It assumes no prior knowledge of data quality control or related subjects.

While data quality theory and methods have advanced, new challenges have emerged in the context of the latest generation of artificial intelligence (AI). A critical issue is the mitigation of the impact of low-quality data through quality control measures.

The book begins by delineating the dimensions and metrics of data quality and the processes for evaluating and analyzing it. It then underscores the impact of low-quality data through various examples. Subsequently, it summarizes the theories at three levels: quality management in data science, pillars of data quality management, and their tools. The recent progress in statistical quality control methods on experimental designs is also briefly discussed. A case study on data quality management in data markets is presented, followed by an exploration of potential research directions involving ghost data and concluding remarks.

The objective of this book is to enhance the capacity of data scientists and engineers in data quality management, particularly within data markets. By doing so, they can create an environment that encourages potential data sellers with high-quality data to join the market, ultimately leading to an improvement in overall data quality.

Chongqing, China   Haiyan Yu
22 Aug 2024

Background and Significance

High-quality data, as a novel factor of production, have assumed a pivotal role in driving digital economic development. The acquisition of such data is particularly crucial for contemporary decision-making models. Data markets facilitate the procurement of high-quality data, serving as data goods, and thereby enhance data supply. This is particularly pertinent in data-scarce domains such as personalized medicine and services. Service personalization leverages user information to tailor services and improve outcomes, with individual data being used to predict user states and requirements for personalization. Data science and artificial intelligence (AI) technologies play a critical role in processing such data. Consequently, potential data sellers with high-quality data are incentivized to enter the market.

In data markets, the challenge for data buyers lies in selecting the most high-quality and valuable data points from various sellers. Controlling data quality is essential to provide solutions for data selection, ensuring that the service meets user requirements. Statistical quality control methodologies often inform the data selection process, such as through experimental design-based quality management. Therefore, data quality management is crucial for the successful operation of data markets.

This book addresses the foundational quality issues in modern data science and data markets. It endeavors to clarify the concept of data quality, its impact on real-world applications (e.g., data exchange and data markets), and the challenges associated with poor data quality. Data scientists have a pivotal role to play in both the intellectual vitality and the practical utility of high-quality data. Moreover, data quality control presents opportunities for data scientists to engage with less structured or ambiguous problems. The book aims to foster fruitful discussions on the contributions that various scientists and engineers can make to data quality. Consequently, the development of data markets stands to benefit from the latest insights and contributions of data scientists in the realm of data quality management.

Acknowledgments

This book originated from the author's time as a visiting scholar in Dennis Lin's laboratory at Penn State University, supported by an international postdoctoral fellowship from the Office of China Postdoc Council (OCPC). During this period, the author participated in research on data quality management and the "ghost data" project initiated by Dennis Lin. The majority of the work was subsequently conducted within the author's research group at the Data and Decision Sciences Center. I extend my sincere appreciation to my colleagues at Penn State University, including Jiayu Peng, Jianbin Chen, Ching-chi Yang, Muzi Zhang, and Nicholas Rios, as well as the students at Chongqing University of Posts and Telecommunications (CQUPT), such as Jiao Xiang and Yali Wang, for their invaluable contributions to the completion of this book.

The publication of this book has been facilitated by the generous support of the National Natural Science Foundation of China (Grants No. 62272077, 72342014), the Chongqing Municipal Science and Technology Bureau (Grants No. 2022TIAD-KPX0155, 2023DBXM008, CSTB2023NSCQ-MSX0073), the Chongqing Municipal Education Commission (Grants No. KJQN202200608, 22SKGH149), and CQUPT.

I am particularly grateful to Robin G. Qiu for his insightful comments, which were pivotal in realizing this work. Additionally, I wish to express my appreciation to the staff at Springer, especially Jialin Yan and Sneha Arunagiri, for their invaluable assistance and support throughout the publication process.

Haiyan Yu

Acknowledgments

Contents

About the Author

Haiyan Yu is an Associate Professor at Chongqing University of Posts and Telecommunications. He obtained his PhD from Tianjin University in 2015. Subsequently, he held the position of Postdoctoral Fellow at the University of Electronic Science and Technology of China from 2016 to 2017 and at Pennsylvania State University from 2017 to 2020. Additionally, he was a Visiting Scholar at Purdue University from 2020 to 2021. His research interests include causal inference and machine learning, personalized medicine, quality management, constrained optimization, and clinical decision support systems.

List of Figures

List of Tables

List of Table

Chapter 1
Introduction of Data Quality Management

1.1 Introduction

1.1.1 Data Quality Issues

Prior to the advent of affordable computing, mainframe computers were utilized by postal courier services to manage name and address data. These computers were employed to rectify name and address data using business rules, identify spelling or typing errors, and track and update customer information, such as a change of address upon relocation. In contrast to manual data correction, this technology offered cost savings for large companies. Companies that provided this service also offered it to others, leveraging the benefits of low-cost computing and server technology to begin controlling data quality (DQ).

Data acquisition measurements are frequently subject to errors, which diminish data reliability and can lead to flawed inferences. Over time, this issue compounds, posing new challenges related to data effectiveness and usability. For instance, a report by the Data Warehouse Institute (Russom, 2006) highlighted that "inconsistent definitions for common terms" are the primary source of data quality issues (see Table 1.1). Data entry errors follow as the second leading cause. Consequently, user interfaces should minimize typing requirements, validate input data before submission, and provide user training to mitigate these errors.

Furthermore, data representing specific business entities, such as customers and products, are particularly vulnerable to quality problems. In organizations, customer data is the most susceptible (accounting for 74% of issues), followed by product data (43%). These entities are more prone to DQ issues than other types of data, such as financial or employee records (Russom, 2006).

DQ issues are typically categorized into seven distinct groups (McKnight, 2013). These categories are:

Table 1.1 Common data quality issues in organizations

Q1: The type of questions ($n = 1522$)	%	Q2: Question source data ($n = 966$)	%
Inconsistent definitions for common terms	75	Customer data	74
Data entry by employees	75	Product data	43
Data migration or conversion projects	46	Financial data	36
Mixed expectations by users	40	Sales contact data	27
External data	38	Data from ERP systems	25
Data entry by customers	26	Employee data	16
System errors	25	Data across a multinational company	12
Changes to source systems	20	Other	10
Other	7		

Note: The data source comprises 1522 responses from 399 respondents in 2005. Q1: Which of the following most often contribute to data quality problems in your organization? Q2: Which types of data are especially susceptible to quality problems in your organization?

1. Admission Quality: This pertains to the correctness of the information being introduced into the system.
2. Process Quality: This involves inspection and quality control measures at each stage of the data processing workflow.
3. Identification Quality: It includes DQ processing techniques, such as record matching and duplicate identification, which can eliminate DQ problems.
4. Integration Quality: This refers to the accuracy of data representation when all related object records are linked and integrated.
5. Usage Quality: It assesses whether the data is used and interpreted correctly when accessed.
6. Timeliness Quality: This category evaluates whether the data remains trustworthy over time.
7. Organizational Quality: This involves the coordination and organization of internal data resources.

Furthermore, service personalization has garnered considerable attention in contemporary service organizations, with practices such as personalized medicine (Wang et al., 2013) leveraging user information to tailor services for improved outcomes (Bonaretti et al., 2020). This approach facilitates the integration of medical resources, services, and health management. Individual data is employed to forecast personalized needs and requirements, with data processing and the integration of AI technologies playing a crucial role (Wang et al., 2013). Consequently, it is imperative to maintain data quality to ensure that the service offerings align with the user's requirements.

The resolution of DQ issues aims to eliminate and mitigate the barriers and adverse effects that data presents when utilized by users (Sebastian-Coleman, 2022). Common strategies for addressing DQ problems include the development of DQ protocols, the reduction of measurement errors, data boundary validation, cross-referencing, modeling, and outlier detection, as well as verifying data integrity. Upon the identification of a DQ problem, its resolution is approached in a systematic manner. A triage process is conducted based on the business impact, the magnitude of the issue, and the number of individuals or systems affected.

The primary factors that are analyzed in the context of DQ problems include (Loshin, 2010):

1. Criticality: the degree to which DQ issues compromise business processes
2. Frequency: the regularity with which DQ problems occur
3. Feasibility of correction: the probability of rectifying the DQ issue
4. Feasibility of prevention: the potential to eliminate the underlying cause of the DQ problem and to implement continuous monitoring

Categorizing DQ problems necessitates a comprehensive assessment of their general attributes and business impact, which will serve as a foundation for determining their severity and prioritization.

1.1.2 Concepts of Data Quality

Data quality (DQ) refers to the attributes and characteristics that data must possess to fulfill specific requirements. A dataset's proximity to reality is indicative of its DQ level; the more accurate and comprehensive the data, the higher its DQ. For instance, in the context of large language models, "hallucinations" can arise due to the influence of data quality on the models' inference and intelligence capabilities (Dziri et al., 2022). Although there are nuances between data and information, we have not delineated a clear distinction between DQ and information quality (Kenett & Shmueli, 2016). In modern applications such as artificial intelligence, which rely on vast amounts of data, data is considered a new form of production and can be traded as a product. However, according to International Business Machines (IBM) data, the annual losses attributable to low-quality data in the United States were approximately $3.1 trillion in 2016 (Redman, 2016). Therefore, improving DQ is pivotal for mitigating the losses incurred by poor-quality data. As a result, various departments are increasingly establishing data governance teams to manage DQ, often as part of their regulatory compliance efforts. Consequently, departments' demands for data extend beyond mere availability to include new DQ requirements.

The US Department of Defense delineated the impact of data quality in its "DQ Guidelines" published in 2003, categorizing it into four distinct areas: prevention, assessment, internal faults, and external faults. These guidelines elucidate the mechanisms, associated costs, and types of data errors (Loshin, 2010). The International Association for Information and Data Quality (IAIDQ) was founded in 2004, offering a platform for professionals and researchers in the field of data quality (Zhou et al., 2013). In 2006, the United States initiated a graduate education program focused on data quality, with support from institutions such as MIT, which contributed to significant advancements in the field (Talburt & Zhou, 2015; Lee et al., 2007). The International Organization for Standardization (ISO) has provided ISO 8000, an international standard for data quality (Aljumaili et al., 2016). This standard is applicable to streaming data, sensor data (Perez-Castillo et al., 2018), and data from smartphones.

China has developed national and industry standards related to data quality, underscoring its significance. For instance, in accordance with the United Nations National Quality Assurance Framework Manual for Official Statistics (DESA), China has established the National Statistical Quality Assurance Framework (2021), which enhances and refines DQ management and improves the quality of statistical data within the country.

Quality control serves as a critical performance indicator of the superiority level in processes or products (Montgomery, 2007). It is a fundamental element of quality management, with a focus on quality specifications (Box & Narasimhan, 2010; Elg et al., 2021). Dr. Edwards Deming encapsulated the quality control system into "14 points," which form the foundation of Deming's quality management theory and serve as the ideological basis for total quality management (TQM) (Deming, 2018b). Additionally, Deming introduced a knowledge system in quality control, emphasizing systematicness, adaptability, epistemology, and psychology (Deming, 2018a).

1.1.3 Structure of This Book

The book is suitable for researchers and graduate students in quality engineering and related quantitative fields, and it assumes no prior exposure to DQ control or associated topics. Chapters 1–4 present foundational material that could collectively form an introductory master's-level course on the topic. A familiarity with quality control and popular statistical modeling approaches at the master's level should provide adequate background for much of the material presented. However, certain sections of Chap. 5 are more mathematical in nature, focusing on experimental designs. In the initial reading, technical arguments within these chapters can be omitted without loss of continuity. Chapters 6 and 7 delve into more specialized topics on data collection in data markets and ghost data, covering recent hot topics and advanced material in data science and data quality management within the context of data markets (Fig. 1.1).

1.2 Brief Overview of Data Quality Management

DQ control has emerged as a prominent topic in interdisciplinary research. Analysis of literature from the Web of Science (WoS) and the China National Knowledge Infrastructure (CNKI) reveals a consistent upward trend in publications relevant to this topic (see Fig. 1.2a). Using keywords such as "data quality," "quality control," "experimental design," "deviation detection," and "intelligent data collection," 845 papers were identified in the CNKI database from 2014 to 2023, with over 50 papers

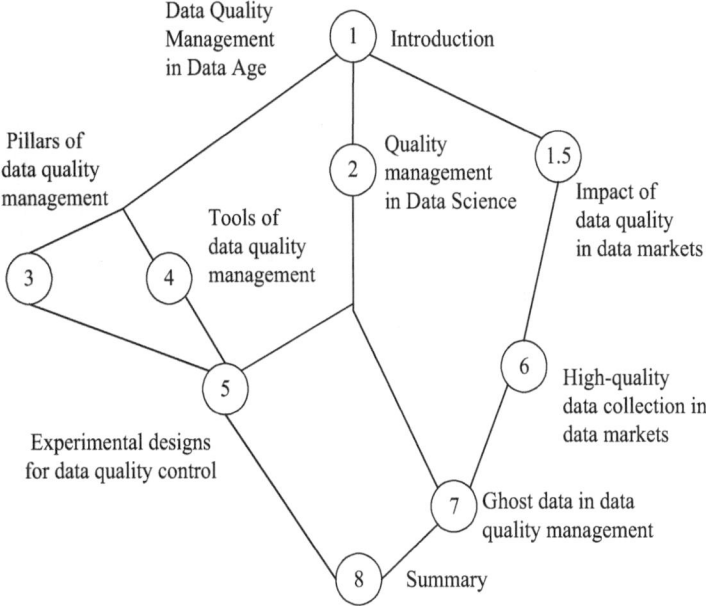

Fig. 1.1 Structure of this book

published annually on average. Of these, 534 papers were related to "data quality," and 346 were related to "experimental design," with some papers featuring these terms as common keywords. On the WoS platform, 116,055 papers were retrieved, with an average annual publication exceeding 10,000 papers over the past 5 years. Notably, 463 papers were related to statistical process control, 206 to supply chain management, 106 to deep learning, and 88 to simulation. This data demonstrates that DQ control, with its interdisciplinary nature, is a significant area of interest from a global perspective.

The top 10 publications on relevant topics (see Fig. 1.2b) indicate that the literature on these subjects in Chinese has remained relatively stable over the past decade. However, there is potential for a more substantial increase in the volume of research outcomes. In contrast, English literature demonstrates a rising trend on these topics, with a more prolific output. On the one hand, the majority of papers that link DQ with experimental design and quality control are the most prevalent, followed by those focusing on optimization, systems, and modeling. On the other hand, Chinese literature displays a relatively higher number of papers that connect DQ with big data and data management.

Among the representative literature, statistical quality control stands out as a pivotal theory and method in quality control, utilizing data analysis to quantify and enhance the scientific nature of quality control processes. These techniques are applied across various fields involving optimization, modeling, and decision-making, enabling accurate inferences based on organized data and informed decision-making under uncertainty. Furthermore, the widespread use of modern

A. Trends of publication

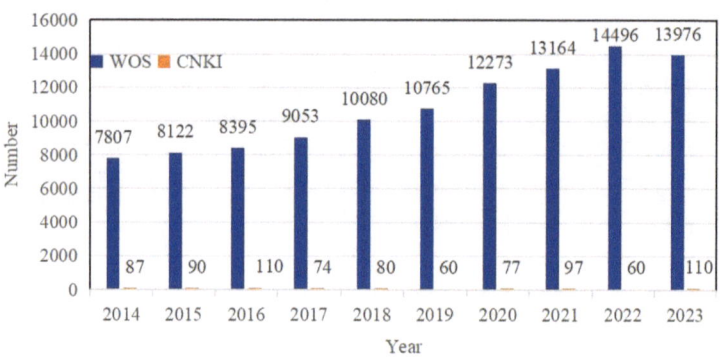

B. CNKI

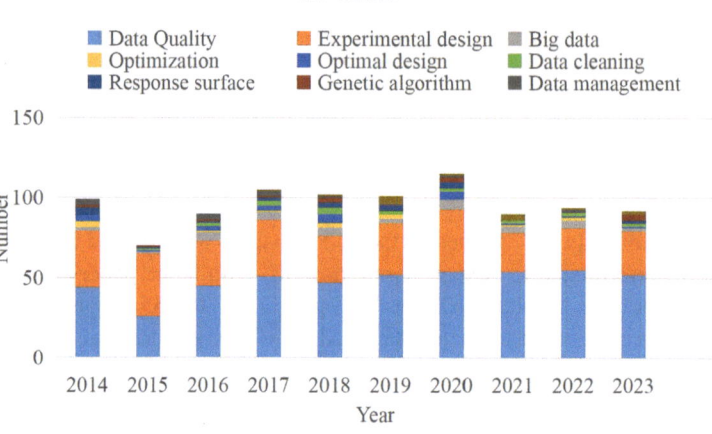

C. WOS

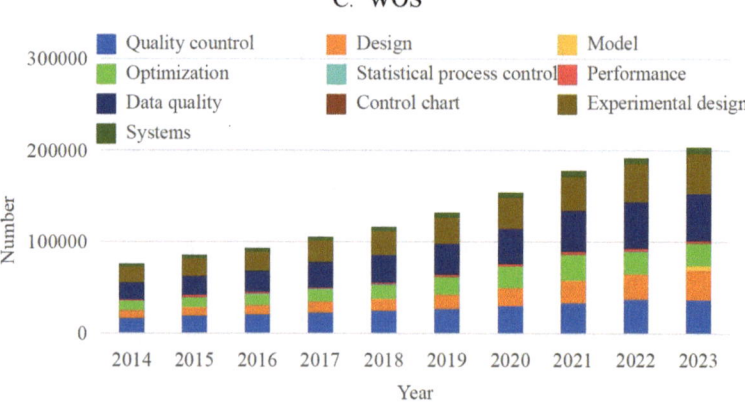

Fig. 1.2 Trend analysis of topics (2014–2023)

computers has expedited the pace of large-scale data computation, giving rise to new demands in DQ control (Efron & Hastie, 2021). Consequently, there is an urgent need for the development of new methods to address these evolving challenges.

Theories and methods concerning DQ have advanced significantly. However, new challenges related to DQ control emerge within the domain of data science. This study makes primary contributions in the following areas:

1. It systematically reviews the pertinent literature on DQ and quality control, summarizing the evolution and recent developments in DQ and its control techniques. The study focuses on works that contribute to DQ theory, highlighting the intricate issues and challenges in DQ and its quality control. Additionally, it outlines representative research frameworks, laying the groundwork for further research on DQ control theory and practical applications.
2. It compares and analyzes DQ from various perspectives, including dimensions, indicators, assessment, and analysis methods. The study's findings, in conjunction with quality control methods, illustrate how to mitigate the impact of low-quality data and enhance DQ.
3. It offers a concise overview of theories, methods, tools, and services for quality control, evaluating their distinctive characteristics, advantages, and disadvantages. Furthermore, it provides a brief review of experimental design methods aimed at enhancing DQ during the data acquisition process.
4. This study assesses the impact of low-quality data through case studies and underscores the importance of DQ control. It also briefly outlines emerging advancements in DQ control, particularly in relation to ghost data, and explores potential research avenues for DQ control within the context of artificial intelligence and personalized services.

1.3 Development of Data Quality Management

Data quality will remain pivotal in the upcoming 5–10 years across various domains, including observational, theoretical, experimental, computational, and data sciences.

1.3.1 Deming and Quality Management

Quality, as a measure of excellence, is a critical indicator of a process or product's worth. The development of quality management has been shaped by numerous contributors. Walter A. Shewhart is renowned as the father of modern quality control, while Dr. Edwards Deming is celebrated for his significant contributions to quality management theory.

Dr. Edwards Deming (October 14, 1900–December 20, 1993) has left an indelible mark on the history of quality management. He began his career at the Washington Census Bureau and later gained fame in Japan, where he was invited by the Japanese Union of Scientists and Engineers (JUSE). From Deming's birth in 1900 to his passing in 1993, and up to 2020, societal lifestyles and research requirements have evolved significantly.

Imagine the items we are accustomed to today, and consider how they would have appeared in 1900. In 1900, "Google" would have meant spending hours sifting through books and materials on library shelves. Social networking in 1900 would have been limited to a physical "like" board on a wall to express preferences. Mobile phones, as we know them, did not exist in 1900, and their applications would have been conceptualized as drawings on stone. Television sets, invented in the 1920s, were not yet available in 1900. By 1993, televisions were generally small and heavy, whereas by 2020, they had transformed into large, lightweight LCD screens.

Concerning communication tools, telephones in 1900 were relatively simple, with a more rounded design and a singular function. By 1993, there was a shift toward enhancing phones with additional functionalities; in 2020, they have evolved into portable mobile devices. In terms of computing equipment, computers in 1900 could be likened to an abacus, while in 1993, they were still clunky and non-portable electronic devices. By 2020, individuals have access to portable electronic products, such as mobile phones and laptops.

Dr. Edwards Deming, a figure from the twentieth century, would have faced significant changes in the domains of "quality management, statistics, and data science" if he were alive today. The following review outlines the development of quality management across three distinct periods: 1900 (the year of Deming's birth), 1993 (the year of Deming's death), and 2020 (the current era).

1.3.2 Progress of Quality Management

Quality, traditionally characterized by fitness for purpose, adherence to specifications, and the pursuit of excellence, has been a concept understood since ancient times. However, the systematic study and definition of quality gained prominence only in the last century, particularly following the Industrial Revolution and the advent of mass production. It became imperative to define and manage quality effectively, as the complexity of manufacturing processes grew. Initially, in the 1920s, the quality objective was to ensure that final products met engineering specifications. As manufacturing processes became more intricate, quality evolved into a discipline focused on controlling process variation to produce products of desired quality.

In the 1950s, the quality profession expanded to encompass quality assurance and quality audit functions. Industries that were pivotal in driving independent quality verification were those where public health and safety were of paramount importance.

In his seminal work, *Out of the Crisis* (Deming, 2018b), Dr. Deming meticulously outlines his "14 points," which form the core philosophy of Deming's quality management theory and serve as the theoretical foundation for total quality management (TQM). His final book posits that quality management necessitates innovative management methodologies and comprehensive knowledge systems. The underlying principles emphasize the importance of systematicness, adaptability to the environment, epistemology, and psychology.

The US National Broadcasting Company (NBC) aired a television special in the 1980s titled *If Japan Can, Why Can't We?*, which introduced Deming to Japan for the first time. His statistical approach led to remarkable success in Japan, where he guided industries to enhance productivity. Consequently, Japan established the Deming Prize in his honor.

DQ assurance and DQ control are integral components of DQ management. Quality control (QC) is typically a subset of quality assurance (QA) activities. Elements within the DQ management system may not be explicitly covered by DQ assurance or control activities and responsibilities but can still involve DQ assurance and control. Figure 1.3 illustrates the definitions of DQ management system from the ISO 9000 standards (Carnerud & Bäckström, 2021).

DQ assurance is characterized as the fulfillment of DQ requirements from a systematic perspective, whereas DQ control is defined in terms of operational techniques and activities. DQ assurance encompasses "a component of DQ management that is focused on ensuring confidence that DQ requirements will be met." The confidence provided by DQ assurance is bifurcated—internally to management and externally to customers, government agencies, regulators, certifiers, and third parties.

DQ control, on the other hand, pertains to "a component of data quality management that is focused on fulfilling DQ requirements." While DQ assurance relates to the execution of a process or the creation of a product, DQ control is more concerned with the inspection aspect of DQ management.

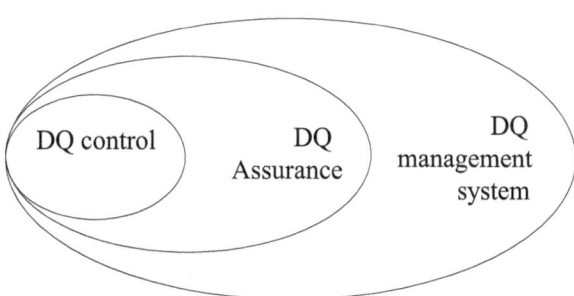

Fig. 1.3 Relationships of DQ management system

1.4   Concept of Data Quality Management

Firstly, I provide an overview of the pertinent definitions of data quality. Subsequently, I delineate the components of DQ control.

1.4.1   Definition of Data Quality

Data quality encompasses a range of definitions, with high-quality data being characterized as data that "fulfills the intended purposes in operations, decision-making, and planning" (see Table 1.2). Historically, data validity has been equated with DQ. Consequently, veridical data science can achieve stability in statistical outcomes through the deployment of predictive models and algorithms (Yu, 2020).

In Table 1.2, the common characteristic of "data quality" is described as the alignment between the actual and desired state of a specific dataset. The desired state is often referred to as "fit for use," "compliant with standards," "meeting consumer expectations," "defect-free," or "meeting requirements." These expectations, standards, and requirements are typically established by domain experts or groups, standard-setting organizations, laws and regulations, business strategies, or software development strategies.

Similarly, information quality (IQ) pertains to the quality of content within an information system. Domain experts may employ more intricate models to define information quality. Although most practitioners do not differentiate IQ from DQ, some researchers ensure that specific contextual quality requirements are met for certain types of information through DQ.

Table 1.2 Definitions of data quality

Perspective	Description	Contributors
User	Data suitable for data users Data that "meets or exceeds consumer expectations" Data that "meets the requirements of its intended use" (Fürber & Fürber, 2016)	Fürber and Fürber (2016)
Business	Data suitable for "use in its intended operations, decision-making, and other processes" or that demonstrate "compliance with set standards" to achieve usage objectives (Herzog et al., 2007)	Herzog et al. (2007)
	Data appropriate for their intended use in operations, decision-making, and planning (Fleckenstein et al., 2018)	Fleckenstein et al. (2018)
	The ability of data to meet the business, system, and technology requirements specified by the enterprise (Mahanti, 2019)	Mahanti (2019)
Standardized	The extent to which a set of intrinsic characteristics of an object (data) (quality dimension) satisfies the requirements	ISO
	The usefulness, accuracy, and correctness of the application of data	NIST

The relevant theoretical frameworks for DQ predominantly encompass four categories:

(a) Applying the principles of statistical process control to DQ in the "zero-defect data" framework (Hansen, 1991)
(b) Constructing DQ frameworks from the perspectives of products (compliance with standards) and services (meeting consumer expectations) (Kahn et al., 2002)
(c) Evaluating data quality based on semiotics in terms of form, meaning, and usage (Price & Shanks, 2005)
(d) Highly theoretical methods for rigorously defining DQ through the nature of information system ontology (Wand & Wang, 1996)

Integrating these theoretical frameworks can provide new insights into the understanding of DQ.

1.4.2 Components of Data Quality Control

Data quality refers to the utility of data usage with a given dataset and analysis method for a specific target question (Kenett & Shmueli, 2016). The process of DQ control can be conceptualized as comprising four components: analysis goal, available data (X), utility, and functions (GXUF).

1. Analysis goal: This component primarily focuses on the 4W1H: what, why, when, where, and how. The analysis target encompasses interpretation, prediction, description, enumeration, analysis, exploration, and verification. For the analysis goal, the primary aim is to enhance its prediction accuracy, fitness, statistical ability, statistical significance, goodness of fit, benefit, reduction of deviation, and bias-variance trade-off, among other metrics.
2. Data availability: Data size and dimensions are primarily dependent on observations and variables. These data sources are relatively diverse and encompass:

   (a) Primary/secondary data
   (b) Observational/experimental data
   (c) Single-source/multi-channel source data
   (d) Instrument collection/protocol data

Furthermore, the data type primarily includes:

   (a) Continuous/classification/hybrid data
   (b) Structured/unstructured/semi-structured data
   (c) Profile/time-series/panel data, and network/geographical data

3. Utility: The utility function can reflect the needs of various stakeholders, including parents, teachers, and policymakers.
4. Functions: Statistical models and methods primarily involve:

(a) Parametric/semi-parametric/non-parametric models
(b) Classical probability/Bayesian theory
(c) Econometric modeling
(d) Data mining algorithms
(e) Probabilistic graphical models
(f) Logistics optimization methods

Thus, the critical issue is how to acquire the data. It primarily includes two aspects:

(a) Data pre-acquisition methods: These methods encompass experimental design, clinical trials, study samples, and computer experiments/simulations. Data acquisition rules include randomization, stratification, grouping, blocking, repetition, sampling design, and design-based records linkage.
(b) Data post-acquisition methods: These methods encompass data processing and cleaning, matching, adjustment, meta-analysis, recovery data, target data cleaning, missing data, outlier detection, re-weighting, combined results, etc.

1.5   Impact of Data Quality in Data Markets

Data markets offer a mechanism for acquiring high-quality data (as data goods) for decision-making models and for augmenting the overall data supply. However, the phenomenon of the "market of lemons" is not immune to the digital age, where low-quality data can be disguised and sold at a premium by masquerading as high-quality items. Consequently, buyers have devised creative strategies to navigate this challenge. Some buyers employ novel techniques to discern high-quality products, while others intentionally utilize low-quality data in unconventional ways.

1.5.1   Impact of DQ on AI Performance

Modern AI applications necessitate substantial volumes of training and testing datasets. This requirement is not only about the availability of data but also encompasses DQ. For instance, incomplete, erroneous, or inappropriate data can lead to the development of unreliable decision models. Trustworthy AI algorithms demand high-quality training and testing datasets, which include metrics such as accuracy, completeness, and consistency. Six DQ dimensions have been identified as impacting the performance of machine learning (ML) algorithms, including classification, regression, and clustering tasks (Budach et al., 2022). To validate the impact on the algorithms, conventional numerical experiments were conducted across three application scenarios, where low-quality data, including both training and testing data, and their mixtures were introduced.

High-quality open-source data is frequently associated with ML algorithms. Numerous studies have examined the DQ of open-source datasets, including Wikipedia, Wikidata, DBpedia, and UCI Machine Learning Databases. DQ analysis for Wikipedia data encompasses entire articles (Mesgari et al., 2015). DQ modeling methods are diverse, including the use of ML algorithms like random forests and support vector machines (SVM) (Hasan Dalip et al., 2009). However, the methods for evaluating DQ differ between Wikidata and DBpedia.

The Electronic Commerce Code Management Association (ECCMA) aims to enhance DQ by developing international standards for data exchange, such as ISO 8000 and ISO 22746 (Färber et al., 2018). ECCMA also builds and maintains a global open standard dictionary for explicitly annotating information, providing a platform for collaboration between DQ and data governance experts. These tagging dictionaries ensure consistent meaning in data transmission and exchange. While these tools and technologies can be utilized to assess general DQ issues and to analyze and verify DQ, they may not be applicable to checking DQ problems in the context of new ML and AI algorithms, such as noisy labels, and imbalanced data (Gupta et al., 2021).

1.5.2 Impact of DQ on Treatment Effects Identification

The proliferation of data from intelligent healthcare systems, wearable technologies, and body area networks presents novel opportunities for identifying medical treatment effects (O'donoghue & Herbert, 2012). Medical treatment effects are metrics employed to assess the comparative outcomes of treatments in medical randomized trials (Miao et al., 2023). They are commonly quantified using the average treatment effect, which calculates the average outcome disparity between individuals allocated to treatment and control groups. Medical treatment effects can be significantly influenced by medical DQ, encompassing factors such as measurement errors, missing data, and confounding by unobserved variables.

Existing open-source tools for assessing DQ often fall short of meeting the required standards, as ensuring the necessary level of granularity in DQ tools involves additional development costs and places significant demands on software architecture. This is also the case in many mobile health (mHealth) applications, electronic health records (EHR), and related software systems. For instance, in mHealth services, mobile devices are frequently utilized for gathering personal activity data, making them susceptible to security vulnerabilities that may result in data breaches. This personal usage can adversely affect DQ, medical treatment effects, patient safety, and confidentiality without adequate security measures.

Consequently, DQ professionals are increasingly seeking software tools to standardize and streamline DQ processes, validate DQ reports, and evaluate underlying data management and indicator reporting systems. For example, collaborations between organizations like the World Health Organization have established unified methods for ensuring DQ across various diseases. However, the use of health mobile devices poses new challenges to health data security and privacy, which directly impact DQ. Despite this, corresponding DQ control theories and tools must be sufficiently developed to address these challenges.

1.5.3   Impact of DQ on Data Exchange and Transaction

In economics, the term "lemons" originated as a slang for defective used cars that are nearly worthless. The concept of a "market for lemons" in the data context refers to the idea that asymmetric information can lead to a decrease in the quality of data goods. In such a data market, data sellers aim to sell all their data items, whether good or bad, and must represent "lemons" as valuable data products. To foster an efficient data market, organizations are increasingly focusing on data management, which includes enhancing DQ, understanding the significance of data, leveraging data to gain a competitive edge, and treating data as an enterprise asset. Some scholars advocate viewing data as a business product (data as a product) (Wang, 1998). Managing data like other products entails managing its entire life cycle. Tools such as information product diagrams can be used to understand and document the process of generating data products, measure them according to specific circumstances, and fundamentally address the DQ issue.

Other studies propose treating data as by-products. By-products are managed differently from products. When data is considered a product, the focus is on the data and its life cycle; when data is treated as a by-product, the emphasis is on managing the data system rather than the data itself. Using the term "data producer" and "data consumer" can be analogous to the manufacturing of products, explaining the data management process. Consequently, DQ is influenced by two interrelated factors: to what extent it meets the expectations of data consumers and how well it serves its intended purpose.

In data exchange, when an enterprise successfully values data, it can be traded as a new factor of production. Empirical research suggests that perceived information quality (PIQ) is a factor of perceived risk and trust belief, which directly affects the efficiency of data exchange (Nicolaou & McKnight, 2006). PIQ predicts trust beliefs and perceived risk, thereby influencing the willingness to exchange data. PIQ is influenced by two critical system design factors: control transparency and outcome feedback (see Fig. 1.4). Control transparency has a significant effect on PIQ (H1), while outcome feedback does not significantly impact control transparency (H2).

To address the data selection challenge in data markets, a federated approach with linear experimental designs has been proposed to minimize prediction errors for machine learning models (Lu et al., 2024). This method does not necessitate labeled validation data and can be optimized through a rapid and federated process. A key advantage of this approach is its compatibility with a decentralized market setting for acquiring data for test set predictions.

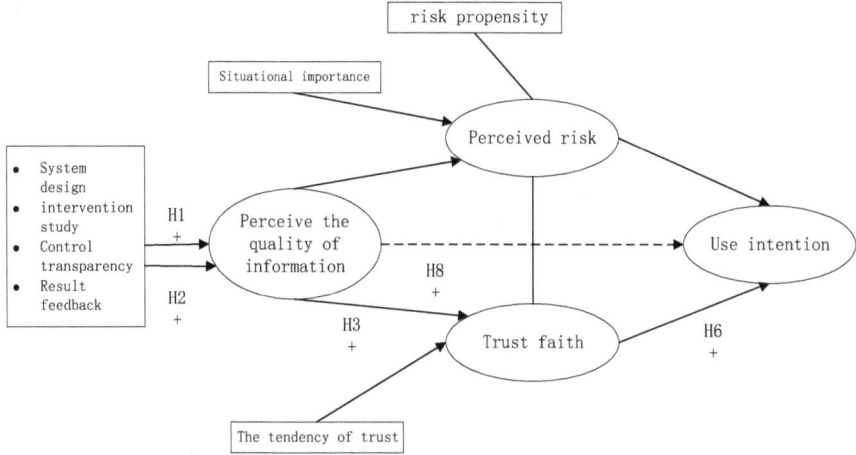

Fig. 1.4 Analysis of factors influencing perceived DQ in data transactions

Therefore, data quality issues, along with a concise overview of data quality management and the influence of data quality on data markets, highlight the importance of acquiring high-quality data. The development of data quality management, in conjunction with the conceptualization of data quality management, illustrates the methodology for identifying the most valuable data points in data markets, thereby ensuring that service offerings correspond with user needs and expectations.

References

Aljumaili, M., Karim, R., & Tretten, P. (2016). Quality of streaming data in condition monitoring using iso 8000. *Current trends in reliability, availability, maintainability and safety: An industry perspective.*

Bonaretti, D., Bartosiak, M., Lui, T.-W., Piccoli, G., & Marchesani, D. (2020). "What can I (S) do for you?": How technology enables service providers to elicit customers' preferences and deliver personalized service. *Information and Management, 57*(6), 103346.

Box, G., & Narasimhan, S. (2010). Rethinking statistics for quality control. *Quality Engineering, 22*(2), 60–72.

Budach, L., Feuerpfeil, M., Ihde, N., Nathansen, A., Noack, N., Patzlaff, H., Naumann, F., & Harmouch, H. (2022). The effects of data quality on machine learning performance. *arXiv preprint arXiv:2207.14529.*

Carnerud, D., & Bäckström, I. (2021). Four decades of research on quality: Summarising, Trendspotting and looking ahead. *Total Quality Management and Business Excellence, 32*(9–10), 1023–1045.

Deming, W. E. (2018a). *The new economics for industry, government, education.* MIT Press.

Deming, W. E. (2018b). *Out of the crisis, reissue.* MIT Press.

Dziri, N., Milton, S., Yu, M., Zaiane, O., & Reddy, S. (2022). On the origin of hallucinations in conversational models: Is it the datasets or the models? *arXiv preprint arXiv:2204.07931.*

Efron, B., & Hastie, T. (2021). *Computer age statistical inference, student edition: Algorithms, evidence, and data science* (Vol. 6). Cambridge University Press.

Elg, M., Birch-Jensen, A., Gremyr, I., Martin, J., & Melin, U. (2021). Digitalisation and quality management: Problems and prospects. *Production Planning and Control, 32*(12), 990–1003.

Färber, M., Bartscherer, F., Menne, C., & Rettinger, A. (2018). Linked data quality of DBpedia, Freebase, OpenCyc, Wikidata, and YAGO. *Semantic Web, 9*(1), 77–129.

Fleckenstein, M., Fellows, L., & Ferrante, K. (2018). *Modern data strategy*. Springer.

Fürber, C., & Fürber, C. (2016). *Semantic technologies*. Springer.

Gupta, N., Patel, H., Afzal, S., Panwar, N., Mittal, R. S., Guttula, S., Jain, A., Nagalapatti, L., Mehta, S., & Hans, S. (2021). Data Quality Toolkit: Automatic assessment of data quality and remediation for machine learning datasets. *arXiv preprint arXiv:2108.05935*.

Hansen, M. D. (1991). *Zero defect data Massachusetts*. Institute of Technology.

Hasan Dalip, D., André Gonçalves, M., Cristo, M., & Calado, P. (2009). Automatic quality assessment of content created collaboratively by web communities: A case study of Wikipedia. *Proceedings of the 9th ACM/IEEE-CS joint conference on digital libraries*.

Herzog, T. N., Scheuren, F. J., & Winkler, W. E. (2007). *Data quality and record linkage techniques* (Vol. 1). Springer.

Kahn, B. K., Strong, D. M., & Wang, R. Y. (2002). Information quality benchmarks: Product and service performance. *Communications of the ACM, 45*(4), 184–192.

Kenett, R. S., & Shmueli, G. (2016). *Information quality: The potential of data and analytics to generate knowledge*. Wiley.

Lee, Y. W., Pierce, E., Talburt, J., Wang, R. Y., & Zhu, H. (2007). A curriculum for a master of science in information quality. *Journal of Information Systems Education, 18*(2), 233.

Loshin, D. (2010). *The practitioner's guide to data quality improvement*. Elsevier.

Lu, C., Huang, B., Karimireddy, S. P., Vepakomma, P., Jordan, M., & Raskar, R. (2024). Data acquisition via experimental design for decentralized data markets. *arXiv preprint arXiv:2403.13893*.

Mahanti, R. (2019). *Data quality: Dimensions, measurement, strategy, management, and governance*. Quality Press.

McKnight, W. (2013). *Information management: Strategies for gaining a competitive advantage with data*. Newnes.

Mesgari, M., Okoli, C., Mehdi, M., Nielsen, F. Å., & Lanamäki, A. (2015). "The sum of all human knowledge": A systematic review of scholarly research on the content of Wikipedia. *Journal of the Association for Information Science and Technology, 66*(2), 219–245.

Miao, W., Hu, W., Ogburn, E. L., & Zhou, X.-H. (2023). Identifying effects of multiple treatments in the presence of unmeasured confounding. *Journal of the American Statistical Association, 118*(543), 1953–1967.

Montgomery, D. C. (2007). *Introduction to statistical quality control*. Wiley.

Nicolaou, A. I., & McKnight, D. H. (2006). Perceived information quality in data exchanges: Effects on risk, trust, and intention to use. *Information Systems Research, 17*(4), 332–351.

O'Donoghue, J., & Herbert, J. (2012). Data management within mHealth environments: Patient sensors, mobile devices, and databases. *Journal of Data and Information Quality (JDIQ), 4*(1), 1–20.

Perez-Castillo, R., Carretero, A. G., Caballero, I., Rodriguez, M., Piattini, M., Mate, A., Kim, S., & Lee, D. (2018). DAQUA-MASS: An ISO 8000-61 based data quality management methodology for sensor data. *Sensors, 18*(9), 3105.

Price, R., & Shanks, G. (2005). A semiotic information quality framework: Development and comparative analysis. *Journal of Information Technology, 20*(2), 88–102.

Redman, T. C. (2016). Bad data costs the US $3 trillion per year. *Harvard Business Review, 22*, 11–18.

Russom, P. (2006). Liability and leverage-a case for data quality. *Information Management, 16*(8), 43.

Sebastian-Coleman, L. (2022). *Meeting the challenges of data quality management*. Academic Press.

Talburt, J. R., & Zhou, Y. (2015). *Entity information life cycle for big data: Master data management and information integration*. Morgan Kaufmann.

Wand, Y., & Wang, R. Y. (1996). Anchoring data quality dimensions in ontological foundations. *Communications of the ACM, 39*(11), 86–95.

Wang, R. Y. (1998). A product perspective on total data quality management. *Communications of the ACM, 41*(2), 58–65.

Wang, P., Ding, Z., Jiang, C., & Zhou, M. (2013). Design and implementation of a web-service-based public-oriented personalized health care platform. *IEEE Transactions on Systems, Man, and Cybernetics: Systems, 43*(4), 941–957.

Yu, B. (2020). Veridical data science. *Proceedings of the 13th international conference on web search and data mining.*

Zhou, Y., Nelson, E., Kobayashi, F., & Talburt, J. R. (2013). A graduate-level course on entity resolution and information quality: A step toward ER education. *Journal of Data and Information Quality (JDIQ), 4*(2), 1–10.

Chapter 2
Quality Management in Data Science

2.1 The Evolution of Quality Management

When viewed from the perspectives of 1900, 1993, and 2020, the development of data science has transitioned from non-existent to nascent and now to its current state. However, more relevant solutions are required.

2.1.1 Development of Statistics

Statistics arises from the fact that things are similar, which enables statistical analysis, and from the fact that things are different, which necessitates statistical analysis (Keller et al., 2017). The classical concept of statistics is tied to information about the state of affairs. The classical conception of statistics is rooted in the gathering of information about the state of affairs. The development of statistics can be traced back to 390 BC, when Yang Shang emphasized the importance of investigation and research, suggesting that "a powerful country knows 13 numbers." These 13 figures encompassed survey data on population, land area, agricultural products, and more, representing the first statistical indicators for a country or region. Consequently, statistical data originated from governmental censuses and research endeavors.

Modern statistics encompasses data collection, organization, analysis, interpretation, and characterization. It often begins with a statistical population or model and has been extensively utilized in scientific research, industrial production, and social life. The statistical community can encompass diverse entities, such as "all people living in a country" or "each atom that constitutes a crystal," and statistics is employed to process data, including planning how to collect data through surveys and experimental design.

© The Author(s), under exclusive license to Springer Nature
Switzerland AG 2024
H. Yu, *Data Quality Management in the Data Age*, SpringerBriefs in Service
Science, https://doi.org/10.1007/978-3-031-71871-7_2

In comparison to the future, the current data scale is still relatively small. There are benchmark-related papers and books that discuss the development of data science, such as *50 Years of Data Science* and *Statistical Inference in the Computer Age*. The latter book focuses on statistical inference, which is closely intertwined with computers, and it foreshadows that new training methods will lead to larger datasets. The book primarily covers classical statistical inference and methods from the early computer age and the twenty-first century, among other related topics. In this context, we focus on strategies like divide and conquer, the power of data, the evolution of statistical theory, and examples of low-quality data.

In contemporary times, statistical methods are applied across all areas that involve decision-making. These methods rely on collated data to make accurate inferences with uncertainty. The advent of modern computers has expedited the pace of large-scale statistical computations, enabling the development of new approaches that would be unattainable manually. Facing new environments, such as analyzing big data challenges, statistics and related disciplines (data science) offer a plethora of opportunities for further vigorous development.

2.1.2   Evolution from Probability to Data Computation

In the realms of statistical research and quality control, there are also notable shifts. The role of the statistician has evolved from a focus on probability theory to a more data-centric approach. Table 2.1 outlines a selection of modern statisticians who have made significant contributions to statistics and quality management.

There are numerous definitions and methods within statistical theory that pertain to probability-based theoretical approaches, and a critical issue is the establishment of the hypothesis. If the hypothesis is flawed, many of the subsequent assumptions can become problematic. It is often necessary to make strong assumptions to derive relevant statistical theories. In contrast, statistics based on data computations place a premium on data quality. Low-quality data can render a model ineffective. All hypotheses must be validated through the model, and the sample data must be confirmed to align with the model's predictions.

Bad Science outlines twelve reasons why scientific endeavors can fail (Goldacre, 2010). The key takeaway is that investigators must carefully design experiments to

Table 2.1 Role of modern representative statisticians (in the past 120 years:1900–2020)

Statistical thinking	Representatives
Probability theory	Karl Pearson (1857–1936) Ronald Fisher (1890–192) Jerzy Neyman (1894–1981) Egon Pearson (1895–1980) C. R. Rao (1919–) George E. P. Box (1919–2013) Edwards Deming (1900–1993)
Data computation	Hadley Wickham (RStudio):COPPS Award (2019) Paul Rosenbaum (Observation Study): Fisher Lecture (2019)

derive accurate conclusions and be diligent in collecting high-quality data during the data acquisition process.

2.1.3   Data Is Power

If Francis Bacon were alive today, it is highly likely that he would assert "data is power" rather than "knowledge is power," as he initially postulated. As a new resource, data possesses the potential and value for application. Consequently, what role does statistics play in modern data science?

The first industrial revolution, which concluded in the late eighteenth century, was a monumental moment in the history of technological development. It ushered in the era of mechanizing manual labor. The second industrial revolution at the turn of the twentieth century saw humanity enter the age of electricity, with significant strides in natural science research. The third industrial revolution began in the 1970s, characterized by the invention and application of atomic energy, computers, space technology, and biotechnology. Many industries underwent transformation, and information technology was revolutionized. The twenty-first century has witnessed a shift in data utilization, referred to as the "digital transformation," primarily in the context of emerging industrial technologies such as Industry 4.0, networking, and the proliferation of massive data and formats. Table 2.2 details the maturity levels of industrial big data analysis.

The innovation theory in quality management primarily encompasses the continuous improvement of the plan–do–check–act (PDCA) cycle, total quality management, and Six Sigma. Its innovation process comprises five stages: definition, measurement, analysis, improvement, and control (DMAIC). Innovative tools utilized in this process include process flow charts, quality-function deployment, fishbone diagrams, control charts, scatter diagrams, and experimental designs.

The challenges associated with traditional tools include:

Table 2.2 Maturity level of industrial big data analysis

Level	Characteristics and limits
Level 1	Dependent on experience while ignoring data
Level 2	Data acquisition on numbers
Level 3	Group the data and make a chart
Level 4	Use census data for descriptive statistics
Level 5	Use sample data for descriptive statistics (association)
Level 6	Using the sample data for inferential statistics (causality)
Level 7	Use real-time sensor data for descriptive statistics and visualization
Level 8	Use real-time sensor data for statistical inference and prediction, decision-making support
Level 9	Based on industrial AI, realize autonomous process control

(a) Process sampling inspection, which only captures a portion of the production or product information
(b) The absence of a data or information fusion platform, leading to difficulties in achieving high-quality data sharing and traceability
(c) A low degree of automation, predominantly reliant on significant human involvement and subjective decision-making

2.1.4 Divide and Conquer

Divide and conquer is a comprehensive data processing methodology, consisting of two primary stages: initially, the large-scale data (matrix) is segmented into blocks (or sub-matrices) (Mackey et al., 2011); subsequently, the results are aggregated using data analysis techniques such as meta-analysis and MapReduce (Yu et al., 2016).

The core challenge lies in determining how to effectively divide, and aggregate, and the specific steps and internal mechanisms. This approach has also been applied to restaurant design during the COVID-19 pandemic, where chefs in Cleveland created custom room dividers and restaurants. The process aligns with Deming's famous dictum that "best efforts are not enough; you have to know what to do"; "what can we do to work smarter, not harder"; and "without data, you're just a person with an opinion."

2.2 Low-Quality Data Examples

From a statistical standpoint, low-quality data poses numerous challenges across healthcare, social life, and engineering domains. This section illustrates four specific instances: two healthcare cases involving COVID-19 testing discrepancies and Google's epidemic trend analysis and two social security cases related to recidivism prediction and Los Angeles' predictive policing.

2.2.1 Detection Bias in COVID-19 Testing

A study previously investigated whether universal testing for COVID-19 is warranted (Yi et al., 2020). In February 2020, Asia, particularly China (see Table 2.3), witnessed the outbreak of COVID-19, which rapidly spread. The disease resulted in a substantial number of infections and deaths. The World Health Organization (WHO) reported all confirmed cases. At that time, the number of confirmed cases in the United States was relatively small.

Table 2.3 COVID-19 cases of Asian countries/regions on February 14, 2020

Country	Confirmed cases	Total deaths	Population (2017)
China	48,548	1381	1.386 billion
Singapore	58	0	5.639 million
Japan	33	1	126.80 million
South Korea	28	0	51.47 million
Malaysia	19	0	31.62 million
Vietnam	16	0	95.54 million
Philippines	3	1	104.90 million
Cambodia	1	0	16.01 million
Thailand	33	0	69.04 million
India	3	0	1.339 billion
Nepal	1	0	29.30 million
Nepal	1	0	21.44 million
Indonesia	0	0	264.00 million
Total	48,741	1383	3.54073 billion

Note: Statistics from Indonesia suggest that its population exceeds that of many countries (except China), yet there were zero reported COVID-19 cases

Assuming that these data have a specific relationship with a country's population, with larger populations generally reporting more confirmed cases, an intriguing observation arises. In Indonesia, at the bottom of Table 2.3, with a population of over 264 million, there were no confirmed cases reported before February 14, 2020. This discrepancy led researchers to question the phenomenon. Some researchers even traveled to Indonesia to investigate the cause. How can we explain this? Information from the Indonesian government indicates that they had not completed their COVID-19 testing at that time. Consequently, we have identified the phenomenon of "COVID-19 detection bias," where the more testing is conducted, the higher the likelihood of detecting cases; in the absence of testing, no data is reported.

Another example is the South Korean "Drive Through Testing (DTT)" data analysis, which revealed a significant increase in testing on February 26. This method allowed individuals to be tested for COVID-19 without leaving their vehicles at 1 of the 500 designated testing sites. The inspectors, wearing protective gear, conducted tests on each person within their vehicle in approximately 10 minutes. Test subjects would receive their results via text message within 3 days. Hundreds of people were tested daily. The statistics of confirmed cases are depicted in Fig. 2.1. The number of cases on February 26 experienced an unusually large surge compared to previous days' data. This spike was due to the launch of the DTT project, not an outbreak of COVID-19. This demonstrates the speed and efficiency of the new data acquisition method while ensuring safety for those being tested.

Similarly, Japan's COVID-19 confirmed data showed a significant jumping phenomenon, as depicted in Fig. 2.2. From January to March 24, the data remained relatively stable. However, it began to surge exponentially on March 24. This date marked a significant jumping point; on this day, Japan announced the postponement of the 2020 Olympic Games to 2021.

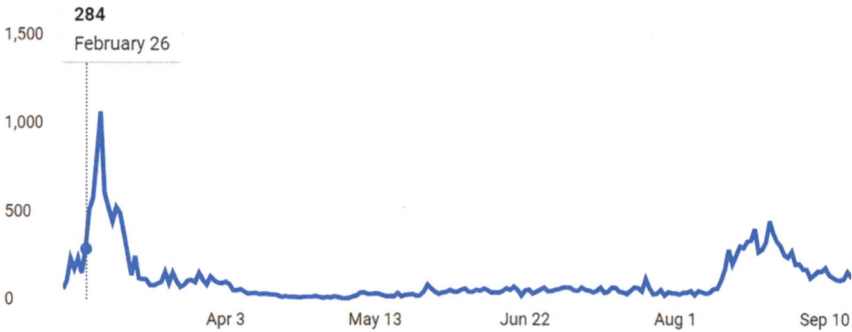

Fig. 2.1 Statistics of South Korea's "Drive Through Testing" (Data source: Google)

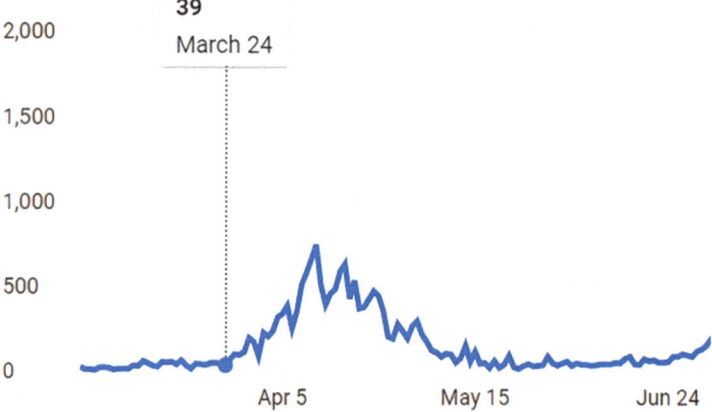

Fig. 2.2 Japan's COVID-19 confirmed data (Data source: Google. On March 24, Japan announced that it postponed the 2020 Olympic Games to 2021)

With this data, how would Deming approach such issues? He might, first, urge the government and the healthcare department to closely collaborate in the collection of high-quality data; second, recommend that the government provide statistics free from political bias and agenda; third, focus on understanding COVID-19 trends through data analysis rather than succumbing to the growing panic surrounding the spread of the virus; and fourth, utilize high-quality data to provide real-time decision support to healthcare policymakers. Furthermore, he would encourage enterprises to break down barriers in various ways and to find new ways to communicate and engage with customers across different departments.

Statistical Thinking (Hoerl & Snee, 2020) also provides fresh insights into these topics. For example, it suggests that the strategic value of data and statistics in addressing practical business challenges should be made evident. It advocates for effective learning principles from both educational and behavioral perspectives and proposes a pedagogical approach to teaching statistics that progresses from concrete examples to abstract theories, moving from the macroscopic (the big picture) to the microscopic (the specifics), and from conceptual understanding to the ability to execute specific tasks. When discussing the integration of computers in statistics, Hoerl and Snee emphasize the role of statistical thinking and methods in problem-solving, encouraging the use of these tools.

More examples of low-quality data can be found across various domains. *Weapons of Math Destruction* (O'neil, 2016) offers insightful commentary. Tracing a person's life trajectory, O'Neil exposes the black-box models that shape our futures by crafting life trajectories. Individuals or societies can make judgments through these models, such as scoring teachers and students, constructing resumes, issuing (or denying) loans, evaluating employees, and mobile health monitoring. O'Neil emphasizes that modelers should take major responsibility for their algorithms and suggests that policymakers regulate their use. Ultimately, it relies on people to govern their mode of life. He also highlights many challenging issues, such as the need for changes in existing unreasonable methods.

2.2.2   Google Epidemic Trend

The Google Flu Trends (GFT) project was initiated in 2008 (Dukic et al., 2012). Its core concept is that when a higher number of individuals are afflicted with the flu, they are more likely to search for flu symptom information on Google. This model aims to simulate and predict the spread of influenza across different regions of the world. The analysis revealed no significant discrepancy between the two influenza trend lines.

The GFT model was intended to assist the Centers for Disease Control and Prevention (CDC) in providing flu forecasts. From 2004 to mid-2009, the prediction trend exhibited a break annually, indicating that the CDC did not release related reports during that period. The GFT was projected to be a continuous trend. For instance, during the H1N1 pandemic in the United States in the spring of 2009, the original GFT model failed to accurately predict the outbreak (Lazer et al., 2014b). Google updated the GFT model in the fall of 2009, but the anticipated accuracy of this monitoring tool remained uncertain.

The influenza prevalence trend, as estimated by four data sources from July 2007 to July 2012, was found to be similar. However, from August 21, 2011, to September 1, 2013, the GFT overestimated influenza prevalence, with its peak surpassing the CDC's estimate by more than double. This discrepancy represents a flaw in the GFT model.

In summary, while the GFT project's motivation was sound, the model and the underlying algorithm were flawed. The GFT failed to detect some critical shifts in influenza prevalence. This machine learning bias led to two primary issues (Lazer et al., 2014a). Firstly, it overestimated the role of big data, suggesting that large volumes of data could generate accurate search terms to measure flu trends.

Secondly, the algorithm lacked an update mechanism, and the GFT required timely adjustments to maintain its accuracy.

2.2.3 Predicting Recidivism

Recidivism refers to the criminal records of individuals who re-offend within 2 years of their release from prison. Predicting recidivism is critical for social security and criminal justice systems. Prediction algorithms are commonly employed to assess the likelihood of a defendant re-offending. These predictions inform pretrial assessments, parole decisions, and sentencing. Proponents argue that big data and advanced machine learning can enhance the accuracy and reduce bias in these analyses compared to human judgments (Dressel & Farid, 2018).

For instance, the COMPAS (Correctional Offender Management Profiling for Alternative Sanctions) software has been used to assess the recidivism risk of over one million defendants. COMPAS employs 137 predictive features and prior criminal records to predict recidivism risk (Dressel & Farid, 2018).

Another ML algorithm used a range of associated variables to predict recidivism rates (Tollenaar & Van der Heijden, 2013). These variables include gender, age, age at first conviction, the most severe crime committed, and its location. When comparing various ML methods, logistic regression was found to outperform other algorithms in predicting recidivism rates, followed by linear discriminant analysis.

However, concerns have arisen regarding the use of such algorithms in predicting recidivism. For example, a black man named Buli Borden, arrested for stealing bicycles and scooters worth approximately $80, was accused of a misdemeanor and had no prior criminal record (Angwin et al., 2016). In contrast, Vernon Pratt, a white man with a history of armed robbery, was arrested for stealing tools worth $86.35. Despite Borden's low recidivism risk prediction and Pratt's high-risk prediction, Borden did not re-offend, while Pratt was sentenced to 8 years in prison for theft.

Why do these ML algorithms fail? For legislators and the legal system, predicting recidivism is a complex task with two levels of risk to consider. The measurement of recidivism is a critical aspect of this issue, and there are ongoing debates about the most effective methods. While ML algorithms can quantify individual crime risks, they may perpetuate human biases. Additionally, these algorithms may not be as accurate as the judgments of ordinary people in predicting recidivism.

2.2.4 Los Angeles Predictive Policing

The objective of predictive policing is to anticipate and prevent criminal activity before it occurs (Ferguson, 2016). This involves leveraging existing criminal patterns to forecast future atrocities and subsequently allocating law enforcement resources to areas where crime is likely to occur. This approach has been adopted in various jurisdictions to combat property and violent crimes.

Whether through predictive algorithms or professional crime analysts, the identification of police patrol zones is a critical component of predictive policing. In a randomized controlled trial conducted by the Los Angeles Police Department (Mohler et al., 2015), the city was divided into multiple 150 m x 150 m grids. The Epidemic Type Aftershock Sequence (ETAS) model was utilized to predict the risk of crime in each grid. The model estimates the probability of a crime occurring in a specific grid n at a particular time t, denoted as $\lambda_n(t)$, as follows:

$$\lambda_n(t) = \mu_n + \sum_{t_n^i < t} \theta\omega e^{-\omega\left(t - t_n^i\right)}$$

where t_n^i is the number of event occurrences in grid n and μ_n is the occurrence rate, which is the non-parametric histogram estimation of the stationary Poisson process. Additionally, $\theta\omega e^{-\omega t}$ represents the triggering kernel of crime data in the model, accounting for near-duplicate or contagious effects (Mohler et al., 2015).

Predictive policing offers several insights:

(a) It is founded on well-intentioned goals.
(b) As with any ML application, predictive strategies can be easily influenced by non-random deviations if not managed carefully. Inaccurate data and feedback can significantly impact predictions.
(c) If ML algorithms are trained on partial data, biased results are likely. For instance, if the training data treats an arrest as a case, the resulting prediction bias might be substantial. Conversely, if the training data treats a crime as a case, the bias is less pronounced.

2.3   Data Quality Requirements and Expectations

DQ requirements clearly articulate the expectations associated with each DQ dimension (Sebastian-Coleman, 2022). DQ measurement needs should illustrate how these expectations are quantified. DQ control experts can synthesize DQ requirements, assessing severity and prioritizing solutions based on their grading experiences (see Table 2.4). Considering the costs associated with the DQ problem and the source of the problem data, by defining cost types (e.g., direct vs. indirect costs), DQ experts can prioritize remediation tasks (Loshin, 2010).

Data quality expectations refer to the rules used to evaluate the validity of data values (Loshin, 2010). Compliance checks for DQs can be measured based on the number of violations in the dataset and the percentage of non-compliant records. When defining DQ metrics, business expectations must be clearly documented, and then data users need to be engaged to clarify their objectives. Most DQ expectations are then translated into rules. For instance, a simple rule might be "Blank fields

Table 2.4 Description of the maturity level of data quality

Maturity	Features	Description
Level 1	Junior	Passively perform DQ activities Failure to recognize DQ expectations DQ expectations were not recorded
Level 2	Repeatable	Limited expectations for certain data issues Expectations related to the internal dimensions of the DQ Associated with the value of the data Identify and report simple errors
Level 3	Well-defined	Identify and document the dimensions of DQ Expectations related to DQ dimensions Expectations related to data values, formatting, and semantics can be expressed using DQ rules The ability to validate data using defined DQ rules Provide a way to assess business impact
Level 4	Regulated	Check and monitor the validity of the data during the process Frequent business impact analysis of data flows Incorporate the results of impact analysis into the prioritization of managing expected compliance Dataset DQ evaluation based on round-robin scheduling
Level 5	Optimized	DQ datums are defined Adherence to DQ expectations related to individual performance goals Industry efficiency levels are used to predict and set improvement targets Control processes enable data validation and integration into business processes

should be filled," or a more complex rule could be "Fill only under certain conditions."

Additionally, DQ expectations are formulated based on the plausibility test (Sebastian-Coleman, 2022). The key to any plausibility check is to set criteria for a particular situation and then assess the actual data to determine if those criteria are met. The criteria include requirements for data content and the data processing process and a deep understanding of the risks and thresholds that DQ poses to the business. The added value of a DQ rule procedure should be measured against its constituent DQ rules in terms of whether they meet business expectations (and, therefore, business value). By evaluating the business impact of low-quality data and the degree to which its causes are linked, the cost-elimination method can be used to assess low-quality data sources.

The data quality expectation check primarily focuses on the points of data usage, examining the consumer expectations associated with each information product. Data users perform backward processing from the perspective of data consumption. At each stage of processing, data values can be created, modified, or left untouched, and they are validated at the end of the processing stage as per the DQ expected value.

2.4   Data Quality in Statistics and Data Science

2.4.1   Statistics and Data Science

Many scholars have contributed to the development of data science. For instance, *Harvard Business Review* (HBR) and Dr. Xiaoli Meng, along with his colleagues, recently launched the *Harvard Data Science Review* (HDSR) journal. Since 2002, Dr. Dennis Lin has been on the *Journal of Data Science* editorial board and has dedicated unremitting efforts to its advancement.

1. Two differences between statistics and AI

   Despite being a foundational discipline for artificial intelligence, statistics and AI serve different roles. The first difference is that AI is used when an answer or solution, whether good or bad, is required; statistics are employed when an answer with an explanation and its reliability or credibility is needed. The second difference stems from our cultural backgrounds: statistics aid in comprehending data, and there is a keen passion for processing, simulating, and analyzing data, among other related activities, whereas computer scientists tend to keep data with codes or algorithms, using computer codes to match data, and hoping that these codes can be applied universally to other data.

2. Low-quality data can deceive

   Big Data, Big Dupe highlights that big data can bring chaos (Few, 2018). It emphasizes the need for clarity and even the absence of widely accepted definitions for big data, even with its 3 Vs, 4 Vs, etc. There is no evidence to suggest that the concept of big data, as described, actually exists. The pursuit of big data diverts attention from our actual need to derive value from data. Many big data claims have led us into a "big data era" that is not entirely rational. The recommendation is to abandon big data and start using data effectively.

3. Intelligent data acquisition for high-quality data

   High-quality data is comparable. Comparing time slots (quarter, month, and day) and customer groups can enhance the understanding of the research objective. For instance, "sales rate this week is higher than last week" is more meaningful than "sales growth rate of 30%." High-quality data can influence behavior. For example, a specific product's production might not increase order volumes but could reduce output. In another example, a coffee shop might be decorated in a style to attract a specific customer base; the design needs to align with customer preferences. The ratio is often used as a data indicator due to its operational nature and straightforward interpretation.

   Intelligent data acquisition is an emerging technology that collects required data using modern intelligent methods and is applied in various fields such as signal processing and detection instruments. It generally involves four steps: filtering, sampling, storage, and processing. It also offers multiple advantages, such as diverse functions and good scalability, ensuring data quality.

2.4.2 Criteria for Data Collection

Randomization and systematicity are standard principles in the data acquisition process. Whether in the natural sciences, social sciences, or engineering disciplines, each research topic has its distinct subject matter, but data acquisition in all fields relies on randomization. Therefore, it is often necessary to verify whether the data acquisition process meets the criteria for randomization and whether the measurement of randomness is sufficiently accurate. Dr. Deming's work at the Census Bureau serves as an exemplar of adhering to such data acquisition standards. With the contributions of many statisticians, including R. A. Fisher, G. E. P. Box, and Dennis K. J. Lin, experimental design has evolved from agricultural statistics to encompass industrial, service, and, more recently, data science applications.

Advances in experimental design also include physical and computer experiments. Specific methods encompass screening designs, supersaturated designs (Lin, 1993) (addressing the small n big p problem), online experimental design, t-covering arrays, Order-of-Addition (OofA) designs, and computer experiment design, among others. The latest computer experiments and simulations include space-filling designs, Latin hypercube designs, and uniform designs. These methodologies of experimental design provide the foundations for the collection of high-quality data.

Therefore, the progression of data science, alongside the establishment of data quality requirements and expectations, and the distinction between data quality in statistics and data science collectively reflect the advancements in data quality theory and methods. Nevertheless, instances of low-quality data underscore the shortcomings in data quality control and the challenges prevalent within the domain of data science.

References

Angwin, J., Larson, J., Mattu, S., & Kirchner, L. (2016). Machine bias. *ProPublica*, May, 23, 2016.

Dressel, J., & Farid, H. (2018). The accuracy, fairness, and limits of predicting recidivism. *Science Advances, 4*(1), eaao5580.

Dukic, V., Lopes, H. F., & Polson, N. G. (2012). Tracking epidemics with Google flu trends data and a state-space SEIR model. *Journal of the American Statistical Association, 107*(500), 1410–1426.

Ferguson, A. G. (2016). Policing predictive policing. *Washington University Law Review, 94*, 1109.

Few, S. (2018). *Big data, big dupe: A little book about a big bunch of nonsense*. Analytics Press.

Goldacre, B. (2010). *Bad science: Quacks, hacks, and big pharma flacks*. McClelland & Stewart.

Hoerl, R. W., & Snee, R. D. (2020). *Statistical thinking: Improving business performance*. Wiley.

Keller, S., Korkmaz, G., Orr, M., Schroeder, A., & Shipp, S. (2017). The evolution of data quality: Understanding the transdisciplinary origins of data quality concepts and approaches. *Annual Review of Statistics and Its Application, 4*, 85–108.

Lazer, D., Kennedy, R., King, G., & Vespignani, A. (2014a). *Google Flu Trends still appears sick: An evaluation of the 2013-2014 flu season*. Available at SSRN 2408560.

Lazer, D., Kennedy, R., King, G., & Vespignani, A. (2014b). The parable of Google Flu: Traps in big data analysis. *Science, 343*(6176), 1203–1205.

Lin, D. K. (1993). A new class of supersaturated designs. *Technometrics, 35*(1), 28–31.

Loshin, D. (2010). *The practitioner's guide to data quality improvement*. Elsevier.

Mackey, L. W., Jordan, M. I., & Talwalkar, A. (2011). Divide-and-conquer matrix factorization. *Advances in Neural Information Processing Systems*.

Mohler, G. O., Short, M. B., Malinowski, S., Johnson, M., Tita, G. E., Bertozzi, A. L., & Brantingham, P. J. (2015). Randomized controlled field trials of predictive policing. *Journal of the American Statistical Association, 110*(512), 1399–1411.

O'Neil, C. (2016). *Weapons of math destruction: How big data increases inequality and threatens democracy*. Broadway Books.

Sebastian-Coleman, L. (2022). *Meeting the challenges of data quality management*. Academic Press.

Tollenaar, N., & Van der Heijden, P. (2013). Which method predicts recidivism best?: A comparison of statistical, machine learning and data mining predictive models. *Journal of the Royal Statistical Society: Series A (Statistics in Society), 176*(2), 565–584.

Yi, G., He, W., Lin, D. K.-J., & Yu, C.-M. (2020). COVID-19: Should we test everyone? *arXiv preprint arXiv:2004.01252*.

Yu, H., Shen, J., & Xu, M. (2016). Resilient parallel similarity-based reasoning for classifying heterogeneous medical cases in MapReduce. *Digital Communications and Networks, 2*(3), 145–150.

Chapter 3
Pillars of Data Quality Management

3.1 Data Quality Dimensions

Data quality dimensions represent a comprehensive set of measurable attributes that data possesses (Coleman, 2013). These dimensions can be utilized to define data quality expectations, simplifying their specification and measurement, and categorizing the types of data quality measurements (Plotkin, 2020). They provide a framework for defining relevant DQ metrics within a specific context and offer a controlled perspective on DQ management. For instance, one definition of data quality dimensions encompasses seven dimensions (Loshin, 2010a): uniqueness, accuracy, consistency, completeness, timeliness, currency (the extent to which data is up to date with the real-world object it represents), and format compliance. Currency, for example, indicates how well data maintains equivalence with the modeled object; despite dynamic changes, the data remains equivalent to the modeled object. Currency can be measured by the expected frequency of updates to master data elements and by verifying the data's currency.

DQ dimensions can also be defined at different levels, such as the ten dimensions defined at two levels (Zhou et al., 2013), comprising four intrinsic and six contextual dimensions. Another definition categorizes DQ dimensions into seven dimensions (Loshin, 2010a): relevance, trustworthiness, data specifications, and so on (see Table 3.1).

DQ dimensions align with business process metrics; DQ metrics are associated with data element values or representations of master data objects. These types of DQ dimensions are highly suitable for system automation, enhancing the effectiveness of data rules within DQ tools.

H. Yu, *Data Quality Management in the Data Age*, SpringerBriefs in Service
Science, https://doi.org/10.1007/978-3-031-71871-7_3

Table 3.1 Dimensions of data quality in representative studies

Level	Dimension	Classification method
1	7	Completeness, validity, accessibility, timeliness, consistency, accuracy (Plotkin, 2020)
1	7	Relevance and trustworthiness, data specifications, foundations of data completeness, accuracy, consistency and synchronization, timeliness, accessibility (McGilvray, 2021)
2	10	Intrinsic level: accuracy, essential content, structural consistency, semantic coherence (Loshin, 2010a)
		Contextual level: completeness, contextual consistency, currency, timeliness, reasonableness, identifiability
4	15	Intrinsic DQ: credibility, accuracy, objectivity, reputation
		Contextual DQ: value-added, relevance, timeliness, completeness, appropriate amount of data (Wand & Wang, 1996)
		Knowledge representational DQ: interpretability, ease consistency, ease of understanding, knowledge representation consistency, concise articulation
		Accessibility DQ: accessibility, access security

3.2   Data Quality Metric

Data quality metrics define the specific data that is being measured and the attributes that are measured (Sebastian-Coleman, 2022). For instance, in healthcare data, a particular metric is employed to quantify the number of effective procedure codes within the main procedure code field of a set of medical records. The metric does not include the threshold itself; it only describes what is being measured. Specific measurements (Measurements) typically use values to describe the state of the data at a particular time. Measurement types establish a bridge between indicators and specific measurements (see Fig. 3.1). A measurement type within the DQ context is a category that enables the repeatable measurement of data that meets the criteria for that type, irrespective of the data content.

In summary, dimensions form the core of DQ, delineating the measurable aspects of DQ and explaining why it is measured (Why). Measurement types are the core of the DQ analysis framework, detailing the general method of measuring dimensions: how to measure (How)—specific descriptions of the data to be measured answer what is being measured. A measurement type is a general form of a particular measurement. In the example of medical diagnostics procedural codes, the measurement type is "validity," which represents the value in the specified column of the database. Data with the same domain can be measured using the same method.

Existing objective data quality (DQ) metrics often cannot be directly applied to evaluate the final performance of the business (Loshin, 2010b) because they do not account for some subjective aspects of DQ and may be influenced by the correlation between data defects within the organization and adverse business effects. This has led to new research directions, such as: How can the impact of DQ problems be identified? How can processes be improved to isolate input from erroneous data sources? How can business value be related to DQ?

Decreasing abstraction.[increasing specificity. Concreteness].Closer proximity to data

Increasing ability to understand and interpret measurement results

Dimension The WHY of measurement	Completeness	Timeliness	Validity	Consistecy	Integrity
Measurement types The HOW of measurement	Compare summarized data in amount fields to summarized amount provided in a control record	Compare actual time of data delivery to scheduled data delivery	Compare values on incoming data to valid values in a defined domain (reference table range, or mathematical rule)	Compare record count distribution of values (column profile) to past instances of data populating the same field	Confirm record level (parent/child) referential integrity between tables to identify parentless child records, (i.e. "orphan") records
Specific DQ metrics The WHAT of measurement	Total dollars on Claim records balances to total on control report	Claim file delivery against time range documented in a service level agreement	Validity of Revenue Codes against Revenue Code table	Percentage distribution of adjustment codes on Claim table consistent with past population of the field	All valid procedure codes are on the procedure code table

Fig. 3.1 DQ dimensions, measurement types, and specific metrics

3.3 Data Quality Assessment and Analysis

To ensure the provision of high-quality data, data quality assessment is a fundamental element of the DQ plan, serving as a reference standard, a basis for requirements, and a method for checking DQ.

3.3.1 Data Quality Assessment

Data quality assessment aims to uncover the correlation between data and DQ expectations and to determine whether the data meets the DQ expectations. The DQ assessment report documents trends, processes, observations, recommendations, and suggestions for identifying and rectifying anomalous sources that lead to critical business exceptions, including methods for identifying and rectifying anomalies (Loshin, 2010b). DQ assessment verifies each indicator and evaluates its consistency and effectiveness in relation to key values and the suitability of DQ measurements for data content and processing. The assessment outcomes are utilized to revise existing measurement and response protocols and to establish new ones (Sebastian-Coleman, 2022). The DQ assessment process identifies data risks and involves reviewing and determining DQ measures with users, ultimately establishing DQ indicators, such as evaluating image quality without a reference standard.

Data quality control methods ("DQ methods") are algorithms that systems can automatically apply to detect and correct DQ problems (Woodall et al., 2014). Large information systems often require appropriate DQ methods to enhance their performance through practical DQ assessment and improvement. Different organizations may have varying requirements for DQ assessment. For instance, some organizations may prioritize compliance with regulations over cost reduction, so they may employ a DQ assessment technique that only partially satisfies their needs and current conditions. One approach to address this is to dynamically configure assessment techniques to achieve optimal application goals using existing assessment techniques.

3.3.2 Maturity Levels of Data Quality Analysis

Prior to commencing data modeling, DQ analysis represents the final stage in the data understanding phase. It involves the examination of the dataset to identify potential flaws, errors, and issues, serving as a general approach for defining DQ rules. Data can be scrutinized to identify patterns, variations, as well as data outliers or anomalies (Herzog et al., 2007).

For DQ expectations, the DQ analysis process typically encompasses the following steps:

Table 3.2 Maturity levels in industrial big data analysis

Maturity	Characteristics and Limitations
Level 1	Use of experience (without using data)
Level 2	Data acquisition (focuses only on numbers)
Level 3	Grouping data and creating charts
Level 4	Use of census data for descriptive statistics
Level 5	Use of sample data for descriptive statistics (associations)
Level 6	Use of sample data for inferential statistics (associations and causal relationships)
Level 7	Use of real-time sensor data for descriptive statistics and visualization
Level 8	Use of real-time sensor data for statistical inference and prediction to support decision-making
Level 9	Use of industrial artificial intelligence for autonomous process control

(a) Determining DQ dimensions
(b) Establishing data granularity (data elements, records, datasets)
(c) Defining DQ constraints and data element dependencies
(d) Documenting constraint conditions and measurement units, selecting measurement methods
(e) Determining acceptable threshold values

Developing a DQ analysis maturity model based on literature analysis and qualitative methods, including interviews with subject matter experts, can enhance the value of data in practical industrial big data analysis. DQ is influenced by maturity levels (Comuzzi & Patel, 2016) (see Table 3.2).

Therefore, data quality dimensions, metrics, and assessment and analysis are the pillars of data quality management. They are fundamental to controlling data quality and ensuring the acquisition of high-quality data for contemporary decision-making models.

References

Coleman, L. (2013). *Measuring data quality for ongoing improvement*. Elsevier.

Comuzzi, M., & Patel, A. (2016). How organisations leverage Big Data: A maturity model. *Industrial Management and Data Systems, 116*(8), 1468–1492.

Herzog, T., Scheuren, F., & Winkler, W. (2007). *Data quality and record linkage techniques*. Springer Science & Business Media.

Loshin, D. (2010a). *Master data management*. Morgan Kaufmann.

Loshin, D. (2010b). *The practitioner's guide to data quality improvement*. Elsevier.

McGilvray, D. (2021). *Executing data quality projects: Ten steps to quality data and trusted information (TM)*. Academic Press.

Plotkin, D. (2020). *Data stewardship: An actionable guide to effective data management and data governance*. Academic Press.

Sebastian-Coleman, L. (2022). *Meeting the challenges of data quality management*. Academic Press.

Wand, Y., & Wang, R. Y. (1996). Anchoring data quality dimensions in ontological foundations. *Communications of the ACM, 39*(11), 86–95.

Woodall, P., Oberhofer, M., & Borek, A. (2014). A classification of data quality assessment and improvement methods. *International Journal of Information Quality, 3*(4), 298.

Zhou, Y., Nelson, E., Kobayashi, F., & Talburt, J. R. (2013). A graduate-level course on entity resolution and information quality: A step toward ER education. *Journal of Data and Information Quality (JDIQ), 4*(2), 1–10.

Chapter 4
Tools of Data Quality Management

4.1 Total Data Quality Management

Data quality control involves managing the use of data by applications. This section provides an overview of data quality control frameworks, plans, and tools.

4.1.1 Uncertainty and DQ Assurance

Data quality measurement necessitates the estimation of measurement uncertainty associated with all analysis results, providing users with the measured uncertainty and its confidence level. Measurement uncertainty can be estimated through a series of procedures, as outlined by ISO (Perez-Castillo et al., 2018). These standards recommend the use of program data, internal quality control data, and efficiency test data based on a component-by-component approach and method validation.

In the presence of these uncertainties, quality assurance (QA) is employed to improve DQ, analyze data, identify anomalies, and perform data cleaning (such as missing data interpolation). For instance, an ML-based DQ assurance framework conducted 2999 quality checks (Sendak et al., 2022). This framework provided 24 quality reports for DQ assurance of five machine learning algorithms across multiple cohorts under various medical conditions, covering 247,536 patient data. DQ assurance is also applied to MRI and other data, with open-source software such as pyfMRIqc provided to enhance DQ and algorithm reliability (Williams & Lindner, 2020).

© The Author(s), under exclusive license to Springer Nature
Switzerland AG 2024
H. Yu, *Data Quality Management in the Data Age*, SpringerBriefs in Service
Science, https://doi.org/10.1007/978-3-031-71871-7_4

4.1.2 *Data Quality Framework*

Data quality control processes are executed both before and after data quality assurance (QA) activities, encompassing the detection and rectification of data inconsistencies. Input data is typically restricted prior to QA. Following QA, statistics are compiled to guide the quality control process, including metrics such as the severity of inconsistencies, incompleteness, accuracy, precision, and the presence of missing or unknown data. Data quality control activities can be integrated into a node within the data quality analysis process (DQ check); they can also be incorporated into DQ estimators for data acquisition and processing at various stages, including peak tables and accuracy assessments. All steps and protocols for data acquisition and processing should be reported alongside the study results. Objective data quality control parameters are evaluated at multiple workflow stages and visually integrated into a graph for direct comparison between preprocessing strategies.

The Framework for Information Quality (FIQ) as proposed in the literature includes (McGilvray, 2021):

(a) The definition of business objectives, strategies, problems, and opportunities
(b) Data life cycle
(c) Key components
(d) Interaction matrix
(e) Positioning and time management
(f) Potential influencing factors

The promotion and application of FIQ theory are encapsulated in a ten-step process:

(a) Defining business requirements and methods
(b) Analyzing the information environment
(c) Evaluating data quality
(d) Assessing business impact
(e) Determining the root cause
(f) Developing improvement plans
(g) Preventing future data errors
(h) Correcting current data errors
(i) Exercising control
(j) Communicating actions and presenting results

These steps are iteratively refined, serving as a mechanism for periodically and continuously evaluating, maintaining, and enhancing data quality.

The DQ control process is characterized by continuous innovation and enhancement of the theoretical framework, evolving from the PDCA loop (plan-do-check-process, plan-do-check-act) to total quality control and then to Six Sigma (Gidey et al., 2014). The innovation process of quality control primarily comprises the five stages of definition, measurement, analysis, improvement, and control, which are also referred to as the DMAIC phases (Li & Al-Refaie, 2008). Innovative tools for quality control include data acquisition flow charts, the House of Quality, control charts, and test design, among others.

However, these traditional DQ control tools still have some limitations, including:

(a) Sampling inspection in the data quality process only captures a subset of the production or product information
(b) The absence of a data or information fusion platform, making it challenging to achieve high-quality data sharing and tracking
(c) A low degree of automation, predominantly dependent on human involvement and subjective judgment

4.1.3 Total Data Quality Management

The concept of total data quality management (TDQM) parallels that of traditional product quality management (Wang & Strong, 1996). Wang et al. conducted a two-stage investigation and ranking of data quality dimensions. The TDQM framework is proposed to encompass four levels (see Fig. 4.1):

(a) Internal DQ, which signifies the quality of data content
(b) Contextual DQ, which stresses the need to consider data quality within the context of the task at hand
(c) Data representation DQ
(d) Accessibility DQ

The latter two dimensions underscore the significance of the system's role. Therefore, high-quality data should be characterized by its content, suitability for

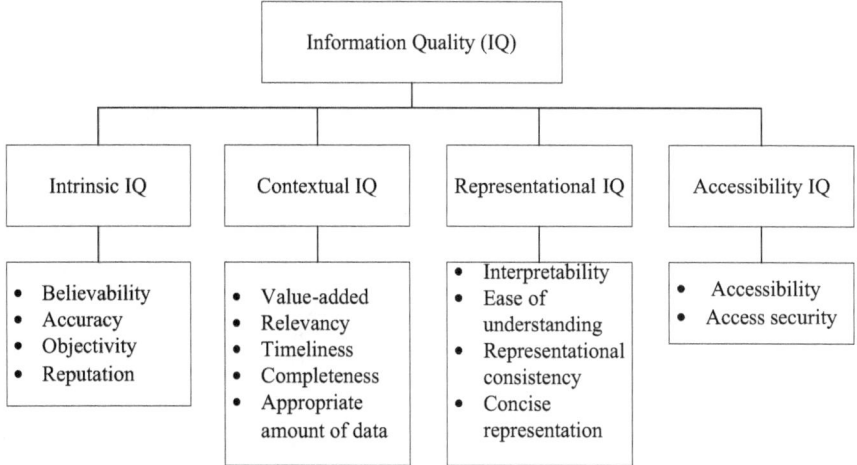

Fig. 4.1 TDQM conceptual framework for data quality

contextual tasks, clarity in presentation, and ease of access for users. In contrast to traditional approaches, TDQM supports database migration, promotes the use of data standards, and employs business rules to enhance DQ.

The TDQM methodology, a step-by-step process, comprises four interconnected links: definition, measurement, analysis, and improvement (Wang, 1998). The first two stages are evaluative, while the latter two phases aim to enhance and develop new tools and methods (Linstedt & Olschimke, 2015). Data quality improvements can be utilized to identify DQ problems, implement process and technical enhancements, enforce data quality standards, and bolster the overall reliability and trustworthiness of an organization's data. The DQ improvement cycle aids in enhancing data quality management, a ten-step process (McGilvray, 2021) (see Fig. 4.2).

This TDQM approach differs from conventional methods in two primary ways:

(a) It emphasizes the use of predefined rules to validate data before it is entered into the database.
(b) It captures and analyzes detected system error data to develop DQ preventive measures during subsequent data acquisition and input.

Nonetheless, research on the critical success factors for effective DQ management remains incomplete. Unstructured data quality remains an under-researched area.

4.1.4 Data Quality Rules

The data quality rule (DQR) assesses data quality at the granularity of the record level, offering a comparative evaluation of the quality of one record against another (Wang et al., 2020). A DQ rule comprises two elements:

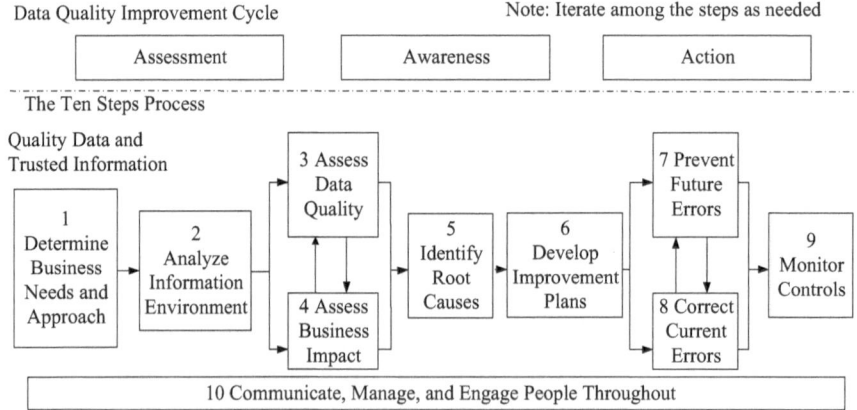

Fig. 4.2 TDQM data quality improvement cycle

(a) The business statement of the rule (business DQR) articulates the meaning of data quality in business terms, describing the significance of the rule to the organization through its application in business processes.
(b) The DQR specification outlines how to verify data quality at the physical level of data storage.

The fundamental essence of DQ rules is to ensure that all necessary rules are defined. At the foundational level, DQRs can be categorized into three types:

(a) Single-column content rules, which verify the content of a single column to ensure it conforms to the specified rules.
(b) Cross-column validation rules, which check the values in multiple columns (typically within a single row of a table) to confirm that the data adheres to the data quality rules.
(c) Cross-table validation rules, which assess the data in multiple tables (and combinations of columns) to ensure compliance with DQRs. Violations of DQRs may also indicate that the rule was incorrectly captured (McGilvray, 2021).

4.2 Data Quality Control Plans and Tools

4.2.1 Data Quality Plans

The Data Quality Program (DQP) offers professional data quality services within a project or operational context (McGilvray, 2021). These services encompass training with specialized knowledge and skills, managing DQ tools, and internal consulting to address DQ issues. They also conduct DQ checks to enhance awareness and develop improvement strategies, among other activities.

One of the objectives of a DQ initiative is to establish a standard process that supports the generation and utilization of high-quality data (Sebastian-Coleman, 2022). The DQP process includes data definition and metadata management, initial data evaluation, ongoing DQ measurements, issue management, and stakeholder communication.

Investigators analyze the DQP from both a management level and a systems approach level. At the management level, the DQP integrates business impact analysis with the expected data representation capabilities of DQ and measures compliance with these expectations. From a systems approach perspective, the DQP establishes standard types of DQ checks to be conducted on the analyzed data sources. This type of activity is typically repeatable and requires fewer specialized resources.

Examples of standard types of data quality checks include:

(a) Integrity of critical attributes
(b) The uniqueness of potential lead candidate entities
(c) The maximum and minimum values of numeric attributes

(d) Frequency distribution of specific attribute types
(e) Pattern analysis of standardized candidate attributes
(f) Cross-source data matching analysis

For instance, to establish a DQP for data quality measurements, a set of metadata (definition of a measure) and data (measurement results) is generated, and its effectiveness is managed and regularly evaluated.

Data quality services represent the operationalization of the DQP (Allen & Cervo, 2015): the implementation of DQ improvements enables organizations to enhance and mature continuously. The DQ service agreement outlines the expectations of data consumers regarding the rules and acceptability of data validity, as well as the reasonable expectation of remediation when data errors and misidentifications occur. For reported incidents, the DQ service agreement defines the resolution goal, including completion and timeline, and may involve multiple steps: addressing one related task at a time and recording the incident in the process record. The DQ management system is updated in a timely manner to reflect the status change.

4.2.2   Data Quality Control Tools

The initial motivation for developing DQ tools stemmed from the need to rectify low-quality data prevalent in database marketing, with early DQ tools primarily focusing on customer name and address cleansing. Commonly utilized DQ tools (see Table 4.1) encompass data profiling, syntactic analysis, standardization, identity resolution, matching record linking and merging, data cleansing, data augmentation, data inspection, and monitoring, among others. For instance, in master data management, the service layer integrates automated processes that handle identity resolution as part of the data acquisition process, rather than batch processing for data cleansing (Allen & Cervo, 2015).

DQ control tools are integral to the data integration process:

(a) Word segmentation and standardization. This involves pattern analysis of data values, identification of value segments, and their integration into canonical data tables.
(b) Data conversion. Here, rules are applied to modify identified error data into an acceptable format.
(c) Record matching. This process evaluates the "similarity" of groups of data instances to determine if they refer to instances of the same master data object.

Furthermore, DQ indicator control tools are categorized into two types: (a) DQ monitors and (b) scorecards or dashboards (Allen & Cervo, 2015). Monitors are employed to detect violations that typically necessitate immediate corrective action. Scorecards assign numerical values to DQ, resembling a real-time snapshot report rather than a real-time monitor. However, scorecards also incorporate monitor reports and can alert the business more promptly to data that does not meet the DQ

Table 4.1 Some examples of data quality tools

Tools	Provider	Description
Data Quality Toolkit	IBM Research (Gupta et al., 2021)	Automated assessment of DQ and remediation of machine learning datasets
Data Quality	R package (Blacketer et al., 2021)	Through a system with over 3300 configurable DQ checks, reporting the observed generality
metabaR	R package (Zinger et al., 2021)	Evaluation and improvement of DNA metabolism encoding DQ
BioGeo	R package (Robertson et al., 2016)	Assessment and improvement of DQ in the occurrence record dataset
Pycheron	Python package (Aur et al., 2021)	Seismic waveform data quality control software package
BUSCO	Python package (Manni et al., 2021)	Assessing genome data quality
pyfMRIqc	Python package (Manni et al., 2021)	Quality assurance for functional magnetic resonance imaging (fMRI)
ODSS	MATLAB macro (Vasta et al., 2017)	Evaluation of microphone sensor data quality
Data Quality Server	SAS package (Svolba, 2012)	Data cleaning on the server for data quality
Visualization tool	JMP package (Karr et al., 2006)	Manual handling and processing of low-dimensional data in a small scale under the guidance of domain knowledge

specifications. DQ monitoring is frequently utilized to depict process behavior, not only to pinpoint anomalies but also to identify the stability and predictability of the final outcome. A high number of data points outside the control limits in a control chart signifies process instability.

Studies have identified that the DQ scorecard encapsulates three aspects of measurement (Plotkin, 2020):

(a) Cumulative DQ scores
(b) DQ strategy, where the data governance team identifies the business impact and defines the corresponding metrics
(c) The level of data governance, which reflects the cumulative score that data stewards should achieve for DQ analysis

In the DQ scorecard, scores are categorized as acceptable (green), risky (yellow), unacceptable (red), or not yet defined (blue). If over 90% of the indicators meet process and threshold criteria, the sample score for the DQ strategy may be green, yellow if between 50% and 90%, and red if below 50%. Rigorous tracking of DQ issues is crucial for conducting risk assessments on project data. In data integration efforts, managing critical stakeholder data is paramount.

In conclusion, software tools that effectively address real DQ issues are considered a fundamental requirement (Karr et al., 2006). Such tools, including system prototypes, can also evaluate and refine new DQ control theories and methods.

Areas that require further in-depth research include algorithmic processing complexity, data scale, and human-computer interaction issues (such as presentation and visualization). Additionally, DQ standards and quality evaluation tools for big data need to be well established (Cai & Zhu, 2015).

4.2.3   Data Quality Control Strategy

DQ control strategies are outlined from the perspective of data acquisition methods to ensure the provision of high-quality data. The development of a data quality strategy (DQS) encompasses:

(a) Assessing the maturity level that aligns with organizational requirements
(b) Evaluating the current maturity level
(c) Documenting any gaps
(d) Establishing the necessary steps to achieve the strategic objectives

Similar to other organizational strategies, DQS must:

(a) Be aligned with the organizational vision and mission
(b) Be integrated into the overall organizational strategy
(c) Evaluate the current status and identify priorities and success criteria
(d) Describe the strategy to achieve the goals
(e) Assign team responsibilities
(f) Establish decision-making criteria
(g) Communicate contributions

The DQ strategy indicators should provide measurable processes, thresholds, and analysis methods that are accepted by customers in key business areas.

The outcomes of DQS further establish DQ system specifications, utilize system software to track operational information, and provide long-term historical quality control charts; middleware can offer more personalized data operations.

Using patient test result data as an example, the functions implemented by this DQ system include:

(a) Automatic validation
(b) Estimating 6σ measurements for all analyses, formulating quality control rules based on 6σ measurements
(c) Automatically turning analyzers on or off based on QC results
(d) Ensuring patient safety from data publication to clinical examination systems and hospital information systems until meeting QC standards
(e) Tracking samples and performing Delta checks for individual patients
(f) Continuously monitoring test turnaround times

Therefore, the deployment of these quality control tools, services, and plans is instrumental in accomplishing the tasks of DQ assurance. They are indispensable for ensuring data quality by comparing actual conditions against requirements and reporting the findings to management.

References

Allen, M., & Cervo, D. (2015). *Multi-domain master data management: Advanced MDM and data governance in practice*. Morgan Kaufmann.

Aur, K. A., Bobeck, J., Alberti, A., & Kay, P. (2021). Pycheron: A python-based seismic waveform data quality control software package. *Seismological Society of America, 92*(5), 3165–3178.

Blacketer, C., Defalco, F. J., Ryan, P. B., & Rijnbeek, P. R. (2021). Increasing trust in real-world evidence through evaluation of observational data quality. *Journal of the American Medical Informatics Association, 28*(10), 2251–2257.

Cai, L., & Zhu, Y. (2015). The challenges of data quality and data quality assessment in the big data era. *Data Science Journal, 14*, 2–2.

Gidey, E., Jilcha, K., Beshah, B., & Kitaw, D. (2014). The plan-do-check-act cycle of value addition. *Industrial Engineering and Management, 3*(124), 2169-0316.1000124.

Gupta, N., Patel, H., Afzal, S., Panwar, N., Mittal, R. S., Guttula, S., Jain, A., Nagalapatti, L., Mehta, S., & Hans, S. (2021). Data Quality Toolkit: Automatic assessment of data quality and remediation for machine learning datasets. *arXiv preprint arXiv:2108.05935*.

Karr, A. F., Sanil, A. P., & Banks, D. L. (2006). Data quality: A statistical perspective. *Statistical Methodology, 3*(2), 137–173.

Li, M. H. C., & Al-Refaie, A. (2008). Improving wooden parts' quality by adopting DMAIC procedure. *Quality and Reliability Engineering International, 24*(3), 351–360.

Linstedt, D., & Olschimke, M. (2015). *Building a scalable data warehouse with data vault 2.0*. Morgan Kaufmann.

Manni, M., Berkeley, M. R., Seppey, M., & Zdobnov, E. M. (2021). BUSCO: Assessing genomic data quality and beyond. *Current Protocols, 1*(12), e323.

McGilvray, D. (2021). *Executing data quality projects: Ten steps to quality data and trusted information (TM)*. Academic Press.

Perez-Castillo, R., Carretero, A. G., Caballero, I., Rodriguez, M., Piattini, M., Mate, A., Kim, S., & Lee, D. (2018). DAQUA-MASS: An ISO 8000-61 based data quality management methodology for sensor data. *Sensors, 18*(9), 3105.

Plotkin, D. (2020). *Data stewardship: An actionable guide to effective data management and data governance*. Academic press.

Robertson, M. P., Visser, V., & Hui, C. (2016). BioGeo: An R package for assessing and improving data quality of occurrence record datasets. *Ecography, 39*(4), 394–401.

Sebastian-Coleman, L. (2022). *Meeting the challenges of data quality management*. Academic Press.

Sendak, M., Sirdeshmukh, G., Ochoa, T., Premo, H., Tang, L., Niederhoffer, K., Reed, S., Deshpande, K., Sterrett, E., & Bauer, M. (2022). *Development and validation of ML-DQA—a machine learning data quality assurance framework for healthcare. Machine learning for healthcare conference*.

Svolba, G. (2012). *Data quality for analytics using SAS*. SAS Institute.

Vasta, R., Crandell, I., Millican, A., House, L., & Smith, E. (2017). Outlier detection for sensor systems (ODSS): A MATLAB macro for evaluating microphone sensor data quality. *Sensors, 17*(10), 2329.

Wang, R. Y. (1998). A product perspective on total data quality management. *Communications of the ACM, 41*(2), 58–65.

Wang, R. Y., & Strong, D. M. (1996). Beyond accuracy: What data quality means to data consumers. *Journal of Management Information Systems, 12*(4), 5–33.

Wang, Z., Talburt, J. R., Wu, N., Dagtas, S., & Zozus, M. N. (2020). A rule-based data quality assessment system for electronic health record data. *Applied Clinical Informatics, 11*(04), 622–634.

Williams, B., & Lindner, M. (2020). pyfMRIqc: A software package for raw fMRI data quality assurance. *Journal of Open Research Software, 8*(1).

Zinger, L., Lionnet, C., Benoiston, A. S., Donald, J., Mercier, C., & Boyer, F. (2021). metabaR: An R package for the evaluation and improvement of DNA metabarcoding data quality. *Methods in Ecology and Evolution, 12*(4), 586–592.

Chapter 5
Experimental Designs for Data Quality Control

5.1 Experimental Designs

The field of statistical quality control is evolving, with a shift in research topics from probability theory to data computation. Statistical theories based on probability theory encompass numerous definitions and methods, with the key challenge being the assumptions required to derive relevant statistical conclusions.

5.1.1 Data Acquisition with Screening Experiments

Data acquisition frequently needs to satisfy two criteria: randomization and systematicity (Box et al., 2005). A typical data acquisition method that meets these criteria is experimental design, which has evolved through agricultural statistics, industrial statistics, service industry statistics, and, more recently, data science. Standard design methods include formal design, informal design, orthogonal design, online experimental design (Bapna & Umyarov, 2015; Yu et al., 2022), sequential design, and others. The advancement of experimental design also includes virtual simulation, computer experimental design, and space-filling design methods such as Latin hypercube design and uniform design. Modern intelligent methods have been employed to collect the necessary data across various fields, such as signal processing and instrument detection. Data acquisition typically involves four stages: filtering, sampling, storage, and processing. Intelligent data acquisition aids in obtaining high-quality data and offers various advantages, such as diverse functionalities and good scalability, which can enhance data quality to a certain extent.

© The Author(s), under exclusive license to Springer Nature
Switzerland AG 2024
H. Yu, *Data Quality Management in the Data Age*, SpringerBriefs in Service
Science, https://doi.org/10.1007/978-3-031-71871-7_5

Screening experiments are designed to identify the most valuable variables among a multitude of candidates. An illustrative example of a response function is shown in Fig. 5.1, which involves input variables from X_1 to X_k.

The screening experiment aims to identify all the valuable variables from X_1 to X_i. The response function can reveal the active factors' (genetic) mechanisms of action. For those variables that do not require explanation, the exclusion process demonstrates how to filter them out. The more factors that are excluded, the fewer will be selected as valuable.

5.1.2   Supersaturated Design

A supersaturated design, also known as a supersaturated factorial design, is a type of experimental design used in statistics. Its purpose is to determine how to utilize $n(<<k)$ observations (experiment runs) to study k parameters. This design is particularly useful when the dataset is small and there is a significant number of potential factors, but only a subset of these factors are expected to have a significant influence.

For example, a technical study of a metal material investigated 24 variables with 14 runs, as shown in Table 5.1. Supersaturated designs are employed when it is impractical or unfeasible to conduct a full factorial design due to the vast number of treatment combinations, which would demand an unfeasibly large number of experimental runs.

In a supersaturated design, the experimenter can still obtain valuable insights into the effects of the treatments, even with a modest number of experimental runs. This is made possible by incorporating a multitude of additional runs with randomly assigned treatments, which assist in estimating the effects of the treatments and their interactions.

Supersaturated designs are a valuable tool for researchers in various fields, such as drug development, agriculture, and materials science, where the need to test a large number of treatments is prevalent. However, it is important to note that while they offer a practical solution to the problem of limited resources, they are not as powerful as full factorial designs for acquiring data to estimate treatment effects. This is because the additional runs with randomly assigned treatments do not

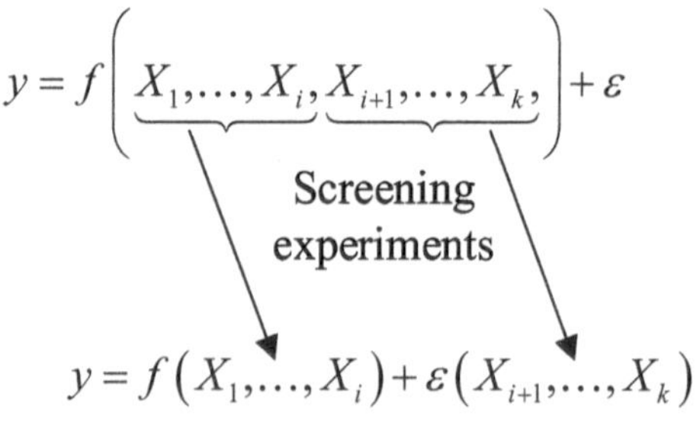

Fig. 5.1 Screening experiment

$$y = f\left(\underbrace{X_1,\ldots,X_i}, \underbrace{X_{i+1},\ldots,X_k}, \right) + \varepsilon$$

Screening experiments

$$y = f\left(X_1,\ldots,X_i\right) + \varepsilon\left(X_{i+1},\ldots,X_k\right)$$

Table 5.1 William's half-fraction experimental design

Run	Factors																								y
	1	2	3	4	5	6	7	8	9	10	11	12	13	14	15	16	17	18	19	20	21	22	23	24	
1	+	+	+	−	−	−	+	+	+	+	+	−	+	−	−	+	+	−	−	+	−	−	−	+	133
2	+	−	−	−	−	−	+	+	+	−	−	−	+	+	+	+		+	−	−	+	+	−	−	62
3	+	−	−	+	+	−	−	−	−	+	−	+	+	+	+	+	+	−	−	−	−	+	+	−	45
4	+	−	−	+	−	+	−	−	−	+	+	−	+	−	+	+	−	+	+	+	−	−	−	−	52
5	−	+	+	+	+	+	+	+	+	−	+	−	−	−	+	+	+	+	−	+	+	+	+	+	56
6	−	+	+	+	+	+	+	−	+	+	+	+	+	+	−	−	+	+	+	+	+	−	−	−	47
7	−	−	−	−	+	−	−	+	−	+	−	+	−	−	−	−	−	−	+	+	+	+	−	+	88
8	−	+	+	−	−	+	+	+	−	+	−	−	−	−	−	−	+	+	+	−	+	+	+	−	193
9	−	−	−	−	−	+	+	+	−	+	+	+	−	+	+	−	+	−	−	−	−	−	+	+	32
10	+	+	+	+	−	+	+	+	+	−	+	+	−	+	+	−	+	+	+	−	+	−	+	+	53
11	−	+	+	+	+	−	−	+	+	−	+	−	−	−	−	−	−	−	+	−	−	−	+	+	276
12	+	−	−	−	+	+	+	−	+	−	+	+	+	−	−	+	−	+	+	+	+	+	+	+	145
13	+	+	+	+	+	−	+	−	+	−	−	+	−	−	−	−	−	−	−	+	+	−	+	−	130
14	−	−	+	−	−	−	−	−	−	−	+	+	−	+	−	−	−	−	−	+	−	+	−	−	127

provide information about the true effects of the treatments, leading to less accurate estimates of the treatment effects.

5.1.3 t-Covering Array

The t-covering array design is commonly employed to assess reliability, particularly in software testing. With modern software, the number of lines of code can be vast. For instance, iPhone applications average 50,000 lines, the Boeing 787 aircraft's control system has 6.5 million lines, the Android system comprises 12 to 15 million lines, the Windows 10 operating system has about 50 million lines, and the Google search engine service has 2 billion lines. A critical code error can lead to system failure. Consider a scenario with 76 input command lines and 276 test cases. If testing is performed at a rate of one million times per second, it would take 2×10^{15} years to complete all tests.

A covering array with intensity t tests all combinations of t-factor levels. The design defines an individual interaction by using a specific set of k factors. The test rule is that if all tests involving the target interaction fail, the interaction indicates a failure. Taking $n = 7$ and $k = 12$ as an example, as shown in Table 5.2, the challenge is to conduct as many tests as possible within a limited time frame. A 2-covering array could study 210 factors in 11 runs, and a 3-covering array could study 25 factors in 23 runs.

5.1.4 Online Experiment Design

As a standard online experiment design method, A/B testing is extensively utilized in Web page optimization. It seeks to derive scientific conclusions through experimental design, sampling, flow segmentation, and testing. The results of A/B testing are considered credible and can be generalized to all online products or services. This approach aids in determining which version of a Web page or application performs better on metrics such as traffic or customer conversion rates.

Table 5.2 A 2-covering array

X1	X2	X3	X4	X5	X6	X7	X8	X9	X10	X11	X12
−1	−1	−1	−1	−1	−1	−1	−1	−1	−1	−1	−1
1	−1	−1	1	−1	1	−1	1	1	−1	1	1
1	1	−1	1	1	−1	1	1	1	1	1	−1
1	1	1	1	1	1	−1	−1	−1	1	−1	1
1	1	1	−1	1	1	1	1	−1	−1	1	−1
−1	1	1	1	−1	1	1	1	1	1	−1	1
−1	−1	1	−1	1	−1	1	−1	1	1	1	1

To conduct an A/B test, an investigator must first create a test page (the experimental group), which may differ from the original page (the control group) in aspects such as title font, background color, and wording. The two pages are then simultaneously assigned to browsing users at random. The investigator subsequently counts and compares the user conversion rates for the two pages, revealing the strengths and weaknesses of the designs. For instance, an online test of two user terminal designs is depicted in Fig. 5.2, showcasing the color changes between the click buttons of the two versions. The determination of which version is superior requires experimentation. By conducting A/B testing, the click conversion rates of the two versions can be quantified and compared. The findings indicate which version has a higher conversion rate, thereby guiding the decision on which version to utilize for the new page design.

The process of A/B testing involves randomly assigning Web site traffic to two different Web designs, collecting data, and performing comparative tests to ascertain if the performance of the two designs statistically differs. This methodology can be extended to compare multiple versions (A/B/n) of Web pages or applications.

Multi-factor A/B testing involves the design and analysis of experiments that integrate factorial designs with A/B testing. This approach arranges, combines, and cross-groups the various levels of two or more factors. The 2×2 factorial design, for instance, encompasses the four combinations of two levels of two factors. The testing process evaluates whether there is an interaction effect between the factors. If such an effect is present, further analysis is conducted to examine the individual effects of each factor. This method can ascertain whether there is an interaction between the factors and identify the optimal combination of online products.

5.1.5 Order-of-Addition Design

The Order-of-Addition (OofA) design is an experimental design methodology employed to determine the optimal sequence of component addition for m components in a process or system (Peng et al., 2019). This approach is particularly useful

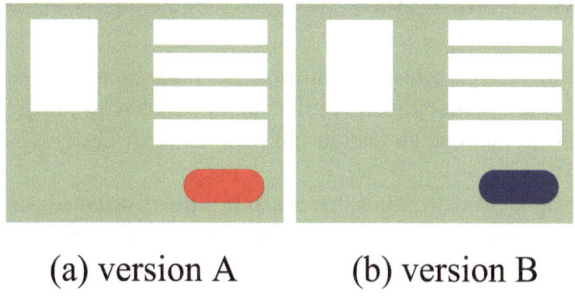

(a) version A   (b) version B

Fig. 5.2 Examples of multi-factor A/B test. (**a**) Version A, (**b**) version B

when there are multiple components or ingredients that can be added in various orders to achieve a desired outcome.

The components in OofA designs are considered permutations, and each permutation is determined by n factorial to ascertain the optimal sequence. For instance, in the context of a culinary recipe, the OofA design might entail testing various sequences of ingredient addition to ascertain which sequence results in the most palatable dish. In Fisher's milk-tea addition experiments (Fisher, 1956), the research question is whether milk added to tea (milk2tea) or tea added to milk (tea2milk) tastes better. The OofA design comprises two combinations of milk2tea and tea2milk, specifically $2! = 2$. In actual experiments, the two varieties may exhibit some repetition.

The OofA design encompasses the following steps:

(a) Identifying the components or ingredients that can be added in different orders
(b) Creating a set of potential sequences in which these components can be added
(c) Conducting experiments where each sequence is tested under identical conditions
(d) Analyzing the results to determine which sequence yields the best outcome

In OofA designs, the number of factors is often greater than 10, with $m! = 10! = 3,628,800$. The challenge in conducting experiments with $m!$ runs is to identify the optimal sequence within a set of n runs. Thus, the OofA design helps quantify the impact of the order of addition on the final product or process, and it can furnish valuable insights into the optimal sequence of component addition.

5.2   Data Acquisition with Experimental Designs

This section provides an overview of designs used in computer experiments and simulations, including space-filling design, L_p star discrepancy, Latin hypercube design (LHD), and its variants.

5.2.1   Space-Filling Design

The space-filling design aims to optimize the placement of n points within a d-dimensional space, with optimality referring to covering as much space as possible. The initial problem setup should consider uniform design (UD), LHD, and Orthogonal LHD.

Uniform designs provide points that are uniformly distributed across the experimental design region (Fang et al., 1982). Fang and Wang first proposed this method in 1982. It is an application of the Quasi-Monte Carlo method in number theory. A literature review (Fang et al., 2000) has systematically examined the theory and industrial applications of uniform designs (Fang et al., 2000).

A Latin square matrix is a square matrix that contains only one sample per row and column. The LHD extends the concept of the Latin square matrix to multidimensional spaces (McKay, 1979). Each Latin hypercube contains at most one sample to each axis perpendicular hyper-plane. Given n (number of runs) and k (number of variables), many Latin squares can be formed. Orthogonal LHD was proposed with the rotation factorial design (Beattie & Lin, 2004). Subsequently, the orthogonal LHD solutions have been optimized (Steinberg & Lin, 2006), and more orthogonal LHD with essential properties have been generated.

5.2.2 L_p Star Discrepancy

The L_p star discrepancy is a metric utilized in experimental design to quantify the uniformity of a set of points across a designated region. This measure evaluates the extent to which a set of points covers a design region, particularly beneficial for high-dimensional spaces. The L_p star discrepancy is calculated as the maximum distance between any point in the design and its nearest neighbor within the design, across all possible lines within the design region.

This metric is based on the L_p norm, which is a method for measuring the distance between two vectors in p-dimensional space. Fang et al. define various uniformity measures with the sequence P (Fang et al., 1982, 2000; Fang & Lin, 2003). The sequence $P = \{x_1, ..., x_n\}$ are points in a multi-dimensional unit cube. Among the various quasi-Monte Carlo methods, the L_p star discrepancy is the most widely used to measure uniformity.

$$D_p(P) = \int_{C^s} \left| \frac{\|P \cap I[0,x]\|}{n} - Vol(I[0,x)) \right|^p dx^{1/p}$$

where $[0,x) = [0,x_1) \times [0,x_2) \times \cdots \times [0,x_s)$, $|P \cap I[0,x)|$ is the number of data points in the sequence P, $Vol(A)$ is a regional volume of A, and $D_p(P)$ is the L_p norm-based discrepancy of sequence P. It is crucial to understand that the L_p star discrepancy is a concept that is distinct from the L_p norm, which is a measure of the distance between two vectors in p-dimensional space. The L_p star discrepancy is a specific application of the L_p norm within the context of experimental design.

5.3 Variants of LHD

The following outlines three LHD variants: orthogonal LHD, sliced LHD, and nested LHD. Sliced LHD and uniform sliced LHD (Chen et al., 2016) are variations of LHD that can be segmented into slices, with each slice maintaining a smaller LHD structure. A derivative method of this is cluster slice LHD, which employs an

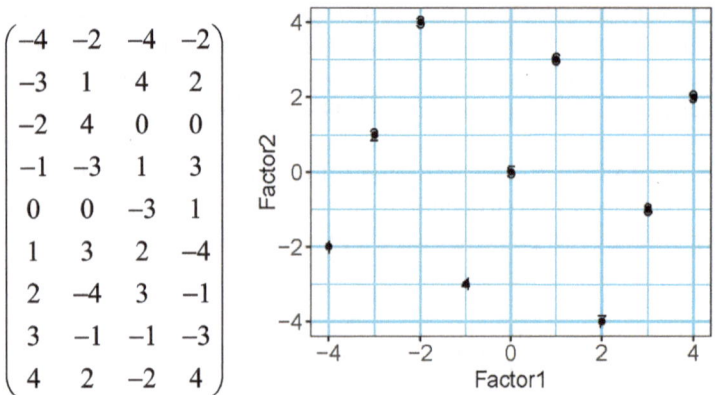

Fig. 5.3 Latin square design L (Beattie & Lin, 2004; Fang et al., 1982)

auxiliary response (AR) to predict the target response (TR). Nested LHD, or nested LHD, can generate computer experiment designs with varying degrees of complexity, ensuring different levels of precision and operational costs. Furthermore, LHD encompasses near orthogonal LHD and Kriging LHD.

5.3.1 Orthogonal LHD

Orthogonal LHD is an advanced form of experimental design that combines the principles of LHD with orthogonality. The key characteristic of LHD is that it provides a near-uniform distribution of the runs across the design space. Orthogonality in experimental design refers to the property where the main effects of the factors are independent of each other. In orthogonal LHD, this orthogonality is achieved by arranging the runs in a way that ensures that the main effects are independent and orthogonal.

Figure 5.3 illustrates an example of a Latin hypercube design L (Beattie & Lin, 2004; Fang et al., 1982) with $n = 9$ and $p = 4$. Each column has 9 numbers from -4 to 4, forming an experimental design with 9 digits. A key characteristic of this design is that the inner product of any two columns is zero, indicating orthogonality. Figure 5.3 (right)[1] deploys the design diagram for the first two columns (Yu & Chen, 2022).

The mirror symmetry orthogonal LHD exhibits second-order orthogonality. An example of such an LHD is depicted in Fig. 5.4. In this design, all main effects are orthogonal, and all interaction effects between any two factors are also orthogonal.

---

[1] The right grid points are those of the first two columns on the left.

Fig. 5.4 Mirror symmetry orthogonal Latin square design

$$OLHD-1$$

$$\begin{pmatrix}
-4 & -2 & -4 & -2 \\
-3 & 1 & 4 & 2 \\
-2 & 4 & 0 & 0 \\
-1 & -3 & 1 & 3 \\
0 & 0 & -3 & 1 \\
1 & 3 & 2 & -4 \\
2 & -4 & 3 & -1 \\
3 & -1 & -1 & -3 \\
4 & 2 & -2 & 4
\end{pmatrix}$$

$$OLHD-2$$

$$\begin{pmatrix}
-4 & -2 & -1 & -3 \\
-3 & 1 & -2 & 4 \\
-2 & 4 & 3 & -1 \\
-1 & -3 & 4 & 2 \\
0 & 0 & 0 & 0 \\
1 & 3 & -4 & -2 \\
2 & -4 & -3 & 1 \\
3 & -1 & 2 & -4 \\
4 & 2 & 1 & 3
\end{pmatrix}$$

Orthogonal LHD is particularly useful in high-dimensional experimental design problems. It provides a way to efficiently explore the design space while maintaining orthogonality, which is important for accurate statistical analysis.

5.3.2 Sliced LHD

Another notable design is the sliced LHD (or SLHD). This design is well suited for computer experiments that involve both qualitative and quantitative factors. The SLHD solution features an ideal slice structure and appealing low-dimensional uniformity. Each slice is also an LHD, maintaining the same low-dimensional uniformity. A new SLHD can be created through symmetric and asymmetric orthogonal arrays, ensuring different uniformities while maintaining the ideal characteristics. The construction method is straightforward to implement and offers significant flexibility in terms of operating scale and the number of factors, distinct from existing approaches.

Figure 5.5 illustrates a special slice LHD, viewed from both macroscopic and microscopic perspectives (Chen et al., 2014). The circles and squares represent two distinct slices. An ideal SLHD should exhibit well-distributed points throughout the design regions in computer experiments.

The uniform SLHD in Fig. 5.6 exemplifies a typical SLHD with circles, squares, and stars representing three different slices (Chen et al., 2016). This design integrates the entire design with the uniformity measurement of each part, ensuring uniformity as a whole. The points of each slice are evenly distributed across the experimental region.

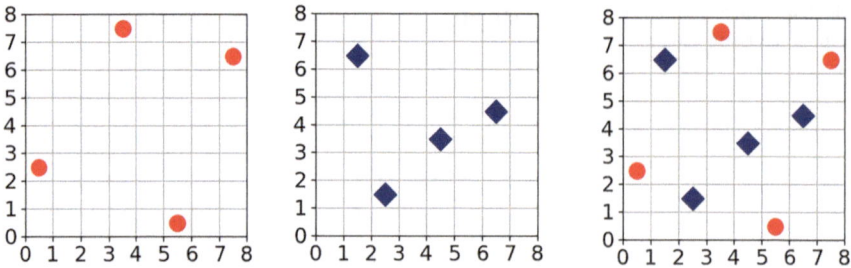

Fig. 5.5 Sliced LHD design (Sliced LHD)

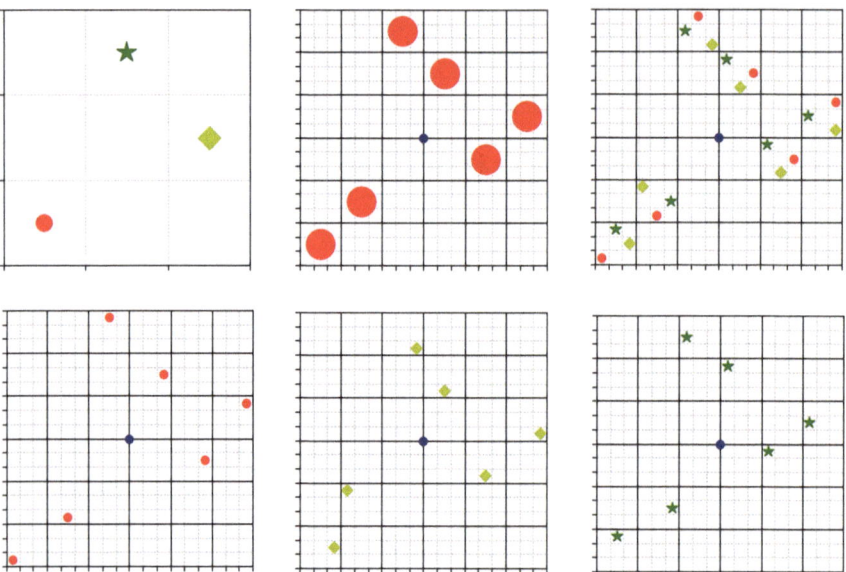

Fig. 5.6 The uniform sliced Latin hypercube design

5.3.3   Nested LHD

The nested LHD (see Fig. 5.7) is designed to facilitate the execution of multiple computer experiments with different levels of precision or fidelity. By employing orthogonal design principles, the experimenter can construct NLHDs with two or more layers. This structured design possesses the following characteristic: the sum of the products of any three columns' elements is zero.

The nested LHD (Fig. 5.7) is well suited for multi-fidelity computer experiments (Yang et al., 2014). It assumes that all factors are continuous and that the design is stratified at the univariate boundary and then becomes uniform at its margins.

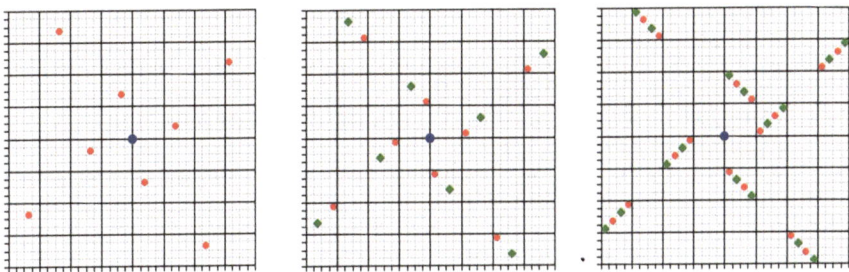

Fig. 5.7 Nested LHD

The nested LHD is particularly useful in situations where the experimenter wants to compare the results of simulations or models with varying levels of complexity or detail. The key characteristic of nested LHD is that it allows for the creation of a hierarchical design where each level of the design represents a different level of fidelity. It also allows for the estimation of the error or uncertainty associated with each level of detail or fidelity.

In summary, experimental designs frequently guide the data selection process. The acquisition of data through experimental designs offers valuable insights into the behavior of the system being investigated. Variants of LHD are robust experimental design techniques that facilitate the comparison of outcomes from simulations or models.

References

Bapna, R., & Umyarov, A. (2015). Do your online friends make you pay? A randomized field experiment on peer influence in online social networks. *Management Science, 61*(8), 1902–1920.

Beattie, S. D., & Lin, D. K. (2004). Rotated factorial designs for computer experiments. *Journal of the Chinese Statistical Association, 42*(4), 431–450.

Box, G. E., Hunter, J. S., & Hunter, W. G. (2005). Statistics for experimenters. In *Wiley series in probability and statistics*. Wiley

Chen, H., Huang, H., Lin, D. K. J., & Liu, M.-Q. (2014). Sliced Latin hypercube designs via orthogonal arrays. *Journal of Statistical Planning and Inference, 149*, 162–171.

Chen, H., Huang, H., Lin, D. K. J., & Liu, M.-Q. (2016). Uniform sliced Latin hypercube designs. *Applied Stochastic Models in Business and Industry, 32*(5), 574–584.

Fang, K., & Lin, D. (2003). Uniform experimental design and its applications in industry. In C. R. Rao & R. Khattree (Eds.), *Handbook of statistics in industry* (Vol. 22). Elsevier.

Fang, K., Wang, D., & Wu, G. (1982). A kind of regression with constraint prescription regression. *Computation Mathematics, 4*(1), 57–69.

Fang, K.-T., Lin, D. K., Winker, P., & Zhang, Y. (2000). Uniform design: Theory and application. *Technometrics, 42*(3), 237–248.

Fisher, R. A. (1956). Mathematics of a lady tasting tea. *The World of Mathematics, 3*, 1512–1521.

McKay, M. (1979). A comparison of three methods for selecting values of input variables in the analysis of output from a computer code. *Technometrics, 21*, 239–245.

Peng, J., Mukerjee, R., & Lin, D. K. (2019). Design of order-of-addition experiments. *Biometrika, 106*(3), 683–694.

Steinberg, D. M., & Lin, D. K. (2006). A construction method for orthogonal Latin hypercube designs. *Biometrika, 93*(2), 279–288.

Yang, J., Min-Qian, L., & Lin, D. K. (2014). Construction of nested orthogonal Latin hypercube designs. *Statistica Sinica, 24*(1), 211–219.

Yu, H., & Chen, J. (2022). Treatment effect identification using two-level designs with partially ignorable missing data. *Information Sciences, 611*, 277–300.

Yu, H., Wang, Y., Yang, C.-C., & Yu, J. (2022). *Martingale stopping rule for Bayesian A/B tests in continuous monitoring. B tests in continuous monitoring.*

Chapter 6
High-Quality Data Collection in Data Markets

6.1 Data as a New Factor of Production

As a new factor of production, data goods have played an essential role in empowering digital economic development and intelligent product-service systems (Demchenko et al., 2018; Ke & Sudhir, 2023). According to the definitions provided in related studies (Demchenko et al., 2018; Spiekermann, 2019), the distinction between data goods and data products is not always clear-cut (Pei, 2020). In this context, we adopt the term "data goods" to encompass data products, related data services, and business applications that utilize data products, as viewed from an economic goods perspective (Demchenko et al., 2018). Data products refer to datasets and the associated information services derived from those datasets (Pei, 2020).

The properties that define viable business and operational models for data goods include data ownership, data quality, and the non-rivalrous nature of data (Jones & Tonetti, 2020). Additionally, data goods possess new economic properties such as sovereignty, trustworthiness, reusability, exchangeability, actionability, and measurability (Demchenko et al., 2018). Consequently, data exchange and trading (DET) mechanisms are established to facilitate data exchange at all stages of the value chain, ranging from data products to data analysis tools and business applications (Fig. 6.1).

The United States leads the global ranking for the number of paid data goods, accounting for 30% of all paid products, followed by Canada (9.3%), the United Kingdom (9.2%), and Germany (7.6%) (Azcoitia et al., 2021). In contrast, the proportion of paid data goods in China is less than 0.5%, but it is growing rapidly. The data factor has been extensively employed across various sectors, including finance, telecommunications, and medicine (Huang et al., 2022b; Yu et al., 2023b). However, there remains a lack of clarity regarding the characteristics of DET modes, such as the specific attributes of data market modes and typical data goods.

H. Yu, *Data Quality Management in the Data Age*, SpringerBriefs in Service Science, https://doi.org/10.1007/978-3-031-71871-7_6

Fig. 6.1 Relationships among data product, services, and data goods

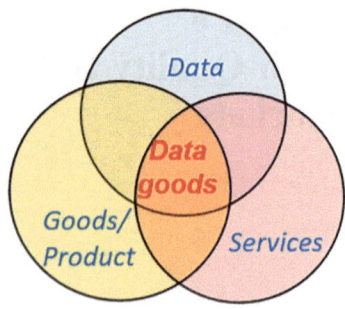

A. Market size of data goods

B. The cumulative number of data exchanges

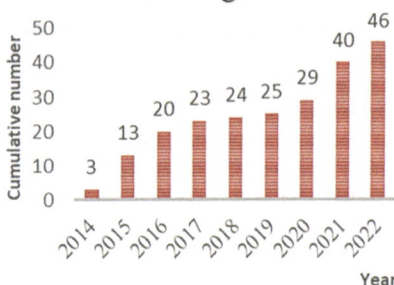

Fig. 6.2 Current trends of data exchange and trading in China. (**a**) The market size and transaction volume of Guiyang Big Data Exchange. (**b**) The cumulative number of data exchanges in China. 2025E stands for 2025 expected. Data source: National Industrial Information Security Development Research Center

The DET of these data forms a sizable data factor market, which is experiencing rapid growth. For instance, the Hainan Provincial government collaborated with China Telecom to establish a data goods supermarket. According to a report by the China Academy of Information and Communications Technology in January 2023 (CAICT, 2023), the market size of China's data factor reached 54.5 billion Yuan in 2020. Projections indicate that the market will exceed 174.9 billion Yuan within the next 5 years (see Fig. 6.2). By October 2022, a total of 30 provinces across China had issued 56 policy documents related to the opening of data markets. Additionally, 193 data platforms (i.e., DET) had been launched by provincial and municipal governments. These platforms offer a range of data services and products, including data queries and data visualization.

The report also indicates that on-site transactions facilitated by data exchanges/ trading centers represent a minor segment of China's data trading market (CAICT, 2023), with the majority of transactions being over the counter, led by companies rather than these centers. The largest data market in China, the Guizhou Big Data Exchange, holds less than 0.2% of the overall market share, with other data exchanges holding even smaller shares. This small market share suggests inefficiencies in the current data exchange model. The report also suggested that on-site transactions dominated by data exchanges/trading centers still were a minor part of China's data trading market (CAICT, 2023), while the major part is over-the-counter transactions led by companies (rather than these centers). For example, the largest data market (Guizhou Big Data Exchange) has only a share of less than 0.2% of the whole data market, and the remaining data exchanges have even less. The small share of the market shows the inefficiency of the data market through the current data exchange.

6.2 Studies of Data Exchange and Trading

This section delineates some scholarly contributions pertinent to DET in the literature, encompassing the types of data goods suitable for exchange and trading, the DET modes, and the efficiency of data markets.

6.2.1 Data Goods for Exchange and Trading

Data has been recognized as a resource, an asset, and a factor of production in recent years. Transaction-level data has been utilized in data-driven research for over a decade (Shen et al., 2020). Research on the DET mechanism during 2010–2015 primarily focused on topics such as trade and capital flow (Sadowski, 2019). Starting from 2018, the research shifted to encompass policy, economic growth, energy management, and smart contracts. Relevant literature has analyzed the exchange and trading processes of data goods from various perspectives, including transaction costs, electronic markets, and data exchange modes, as well as the impact of data monetization on data transactions.

Moreover, data services can be regarded as digital goods (Jiao et al., 2017) or even as knowledge goods (Ba et al., 2001). Data goods exhibit several new characteristics compared to conventional goods, such as value asymmetry, non-rivalry, uneven data quality (Yu & Chen, 2022), and security compliance. The limitations of data exchanges are multifaceted, including the relatively small trading volume of on-site transactions in comparison to the vast trading volume of over-the-counter transactions, market inefficiencies due to asymmetric information, and the high cost of computing power required for robust privacy protection.

Despite numerous academic contributions, there is a scarcity of studies that provide a uniform definition of "data marketplaces" (Spiekermann, 2019). A systematic literature review of data marketplaces examined 133 academic articles (Abbas et al., 2021), indicating that data market-related studies are predominantly technical in nature, focusing on topics such as auction pricing and architecture (Rasouli & Jordan, 2021). These represent the initial stage of data market research (i.e., platform design) (Kolaitis, 2018). However, to progress to the next stage (i.e., platform

adoption), there is a dearth of empirical studies, including topics such as the classification of data markets and their market models.

6.2.2   Modes of Data Exchange and Trading

The United States boasts the most diverse modes of DET (Taylor et al., 2022). Their
market policies are notably open (Ascarza et al., 2021). The establishment of numerous comprehensive data exchange centers in the United States, such as BDEX and
RapidAPI (Kingaby, 2022), is a testament to this. Furthermore, many data traders in
the United States specialize in specific domains, including Quandl and Qlik Data in
economics and finance, Factual in location data, and GE Predix in industrial data
(Azcoitia & Laoutaris, 2022).

The European Union (EU) places significant emphasis on the top-level design of
data legislation (Kozyreva et al., 2021), having bolstered the construction of data
sovereignty. Germany, for instance, has spearheaded the creation of a data space and
the establishment of an exchange and trading system (Heidorn & Weche, 2021). In
the United Kingdom, the financial sector was the first to foster a data market, with
other sectors such as real estate now following suit (Treleaven et al., 2021). The EU
has also championed the Digital Single Market Strategy (Thouvenin et al., 2021),
which aims to provide competitive advantage for their enterprises in the digital
economy.

Data-driven transactions between enterprises are predicated on the exchange and
trading of data. Various electronic transaction modes have emerged in data markets,
such as JD.com's provision of transaction-level data (Shen et al., 2020). The
e-commerce market can be categorized into several types (Qin et al., 2021), including the data pipeline (1 to 1) model and the customer-led data mart (n to 1) model
(Tian et al., 2018). However, these e-commerce models do not adequately address
data-driven transactions due to data's unique economic characteristics, such as
reuse, high fixed costs, low replication costs, non-rivalry, and dependence on specific scenarios or algorithms (Agarwal et al., 2019). Despite advancements, the
characteristics of the data marketplace still lack empirical evidence, distinguishing
it from conventional e-commerce platforms.

6.2.3   Efficiency of Data Markets

Existing research on DET theory primarily focuses on two areas: data production,
such as the repurposing of data for enterprise operations (Yu et al., 2023a), and data
monetization, which includes strategies to enhance data value and ensure its security. The efficiency of resource allocation can be gauged by competitive markets
(Udry, 1996). The Herfindahl-Hirschman Index (HHI) is commonly used to assess
the competitiveness of factor markets (Li et al., 2023). A market is deemed

competitive when its HHI is below 1500, moderately concentrated with an HHI between 1500 and 2500, and highly concentrated with an HHI above 2500. Estimating the HHI for data markets is complex, as the market share of each firm must be determined, and the major data trading occurs through over-the-counter transactions.

In terms of data reuse, data users can submit suboptimal models and still receive an updated, high-quality model (Haupt & Mugunthan, 2021). Auction theory is employed to optimize pricing and the allocation of digital goods. However, data suppliers may provide outdated information or withdraw without proper incentives, leading to inefficiency in the data market. To enhance the value of data, data goods can be systematically monetized for trading (Spiekermann, 2019). The inefficiency of the current data market cannot be fully explained by conventional economic allocation theories, such as the Coase theorem, which relies on well-defined property rights and feasible negotiations. Since data property rights are often not clearly assigned and negotiations may be impractical for each customer, the Coase theorem does not adequately account for the externalities of data goods.

On the issue of data security, research on the influence of privacy-related factors suggests that Chinese consumers are willing to compromise their privacy for corresponding benefits (Liu et al., 2021). Consequently, many firms with extensive consumer data are actively engaged in DET. However, the aggregation of large-scale data has given rise to the data monopoly dilemma. Despite these advancements, determining which data market data goods suppliers (DS) should opt for remains an unresolved issue. This book delves into these research questions and provides empirical insights into the characteristics of data market modes.

6.3 Data Market Model

This section presents a data market model that takes into account the interrelationship between data goods suppliers and demanders. The data market model elucidates the characteristics of data goods and the underlying modes (see Fig. 6.3).

6.3.1 Data Goods Trading Modes

With a factor design that categorizes suppliers $(1, n)$, exchange $(0, x)$, and demanders $(1, m)$, the DET modes can be classified into five distinct categories (M1 to M5) (Cappiello et al., 2020; Huang et al., 2022a). In this design, m (or n) represents multiple subjects, and x signifies either one or multiple subjects. The five DET modes articulate the entity-relationship between data goods suppliers and demanders (see Table 6.1 and Fig. 6.4).

These modes can be categorized into five classes: pipeline, customer-led, supplier-led, data platform, and market-maker. The pipeline mode (M1: 1-to-1)

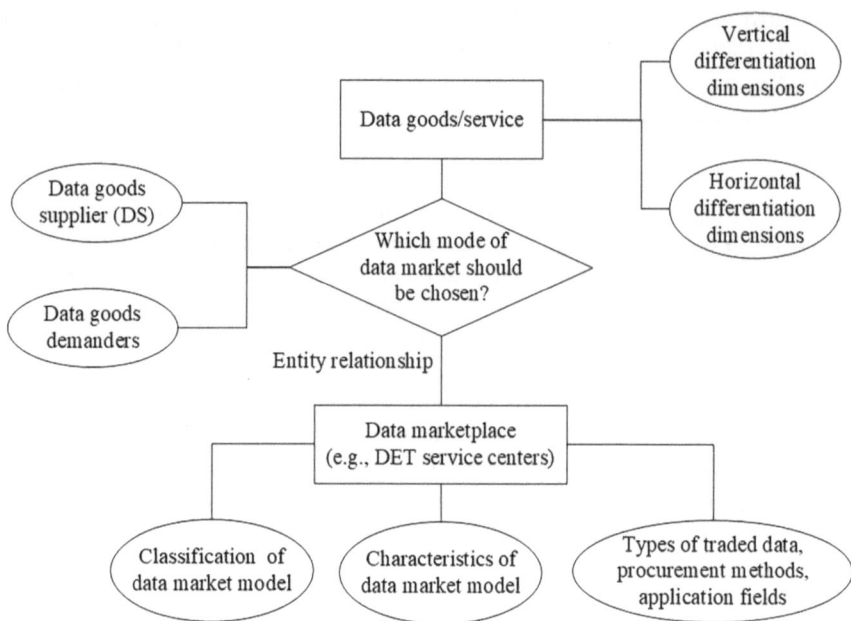

Fig. 6.3 Framework of the data market model

Table 6.1 Market classification for DET in terms of entity relationship

	Supplier		Exchange		Demander			
	1	n	0	x:1/n	1	m	Mode	Descriptions with examples
M1	√		√		√		Pipeline (1-to-1)	Direct exchange and trading resource swap
M2	√		√			√	Customer-led (1-to-n)	App service for user's data API service; membership service
M3		√	√		√		Supplier-led (n-to-1)	Data cloud services
M4		√	√			√	Data Platform (n-to-m)	Data supply/demand aggregation under regulation
M5		√		√		√	Market-maker (n-to-x-to-m)	Contacted in a market under supervision with data protection technology

involves a single supplier and a single demander without an exchange (Deshmukh et al., 2020). For instance, a design choice for a query plan might involve transferring data between two operators in the context of designing query processing primitives.

The customer-led mode of the data market (M2: 1-to-n) can be implemented in various scenarios (Slater & Narver, 1999), such as Baidu App services for users'

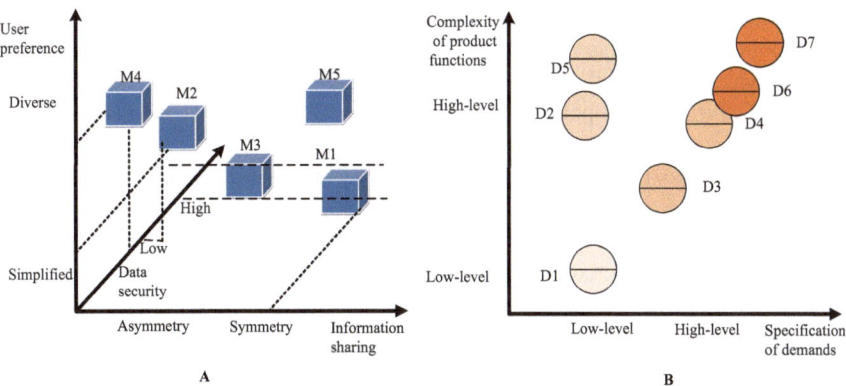

Fig. 6.4 Illustrative diagrams describing the characteristics. (**a**) Data market modes; (**b**) data goods

data, authorized data access from vendors through API, and membership services for data access.

The supplier-led mode of the data market (M3: *m*-to-1) can be implemented in scenarios like data cloud services. The data platform mode of the data market (M4: *m*-to-*n*) can be implemented in scenarios involving data supply/demand aggregation under regulation. The market-maker mode of the data market (M5: *m*-to-*x*-to-*n*) can be implemented in scenarios such as the Guiyang Big Data Exchange (Carvalho, 2020).

In summary, M5 is the most likely mode where the market-maker does not hold data but provides virtual bridges for DET between supply and demand parties. Considering the entity relationship between Data Suppliers (DS) and demanders, the mode of the data market can be recommended for the DS based on further consideration of their specific characteristics.

These data goods can be characterized by vertical and horizontal differentiation dimensions (see Fig. 6.4) (Lauga & Ofek, 2011). The vertical differentiation dimension encompasses features that characterize data goods, utilizing objective measurements (e.g., the complexity of the goods) to distinguish their functions (Lauga & Ofek, 2011). The horizontal differentiation dimension represents features that characterize the reviews of data demanders using objective measurements, i.e., the specification of their demands (Wattal et al., 2009). These two dimensions categorize the data goods into seven classes from D1 to D7.

6.3.2 Characteristics of Data Market Modes

Characteristics of data market models can be encapsulated in three key aspects: user preferences of data demanders (Panico & Cennamo, 2022), data security in information sharing (Liu et al., 2021), and information symmetry in contract design (Zhu et al., 2022). User preferences in data markets pertain to the opportunity to select data demanders. Data security encompasses the process of protecting data goods throughout their life cycle in data markets, shielding them from corruption or unauthorized access. Information asymmetry in the contract design of data trading arises from the fact that data suppliers (DS) are unaware of the private information or the preferences of data demanders (see Table 6.2).

Mode M1 (1-to-1) is typically characterized by simplified user preferences, high data security, and information symmetry in sharing between stakeholders, with a low level of market supervision. This mode is suitable for direct DET. The strengths of this class include its credibility for offline transactions and its control and protection against third-party interference. However, its weaknesses lie in the requirement for increased transparency to facilitate supervision and its vulnerability to the interests of third-party data subjects.

Mode M2 (1-to-n) frequently involves diverse preferences, low security, and asymmetry in information sharing, with limited transparency for supervision. This mode offers more opportunities for DET than the pipeline mode, allowing for data

Table 6.2 Characteristics of data market modes

	Mode	Characteristics	Strengths	Weaknesses
M1	Pipeline (1-to-1)	Simplified preference, high security, information symmetry of sharing, not transparent	(1) Credible for offline transactions (2) Protected from third-party encroachment	(1) Not transparent for supervision (2) Infrangible to third party
M2	Customer led (1-to-n)	Diverse preference, low security, asymmetry of information sharing, not transparent	(1) More opportunities for DET (2) Guaranteeing subjects' interests (3) Data cooperative exploration	(1) Risks of data abuse/privacy (2) API development capability
M3	Supplier led (n-to-1)	Simplified preference, high security, asymmetry of information sharing, transparent	(1) Conducive for development of professional data (2) Transparency for supervision	(1) High standard to SME/customers (2) Increasing digital gap (3) Data monopoly
M4	Data platform (n-to-m)	Diverse preference, low security, asymmetry of information sharing, transparent	(1) Opportunities for data aggregation (2) Transparency for regulation (3) Protected rights of data.	(1) Intercepted by the exchange (2) Limited range of data
M5	Market-maker (n-to-x-to-m)	Diverse preferences, high security, asymmetry of information sharing, transparent	(1) Promoting computing transactions (2) Intensification and specialization of data analysis (3) Transparency for supervision	(1) High cost (2) May be low efficiency (3) Intercepted by the exchange

cooperative exploration. However, it also carries the risk of data theft by unauthorized parties.

Mode M3 (*m*-to-1) often entails simplified preferences, high security, and asymmetry in information sharing. The strengths of this mode include its support for the development of professional data and its transparency for data supervision. However, its weaknesses are that it can be susceptible to data monopolies and that it exacerbates the digital divide between data suppliers and demanders.

Mode M4 (*m*-to-*n*) often encompasses diverse preferences, low security, asymmetry in information sharing, and transparency for regulatory oversight. Its strength lies in its ability to protect data rights. However, its weaknesses are that it can be intercepted by exchanges and involves a limited range of data.

Mode M5 (*m*-to-*x*-to-*n*) makes raw data accessible while maintaining invisibility, ensures controllable data goods that are open to monetization, and enables credible and traceable DET. This mode is characterized by diverse preferences, high security, asymmetry in information sharing, and transparency for supervision. Its strengths include its facilitation of DET and support for regulation. However, its weaknesses are that it can be high cost or inefficient at the initial stage of market development.

6.3.3 Characteristics of Typical Data Goods

According to a survey of China's data market (Huang et al., 2022a), data goods can be classified into seven categories. These data goods encompass permit service (D1), Application Program Interface (API) (D2), datasets (D3), data processing service (D4), data analysis tool services (D5), data application service (D6), and a combination of data products and services (D7) (see Table 6.3). To supplement the dimension of asset specificity, we introduce the specification of demands to articulate the characteristics of typical data goods.

In terms of the specification of demands (Panico & Cennamo, 2022), if data goods are stand-alone, their attractiveness can satisfy the customer's baseline demand. Considering the price P and innovativeness T, the specification of the demand (D) can be expressed as:

$$D = \left(T \cdot P^{-\sigma} \right)^{\omega}, \quad \omega > 0$$

where the parameter σ represents the price elasticity and ω captures the importance that users attribute to the attractiveness of the data goods. The specification of data goods accounts for the size effect without factoring in the users' actual benefits (from the duplication of data). The complexity of goods can be measured using various methods, such as the number of skilled tasks involved in production (Rodríguez-Crespo & Martínez-Zarzoso, 2019) or the number, diversity, and interrelatedness of product variants and components (Trattner et al., 2019). The complexity of data

Table 6.3 Characteristics of typical data goods

Type	Data goods	Complexity of goods	Specification of demands
D1	Permit service	Providing unified codes, module classification, data table, or Structure	Low price elasticity, low innovativeness
D2	API	Applications to interact with a database management system to retrieve and/or update data	Low price elasticity, medium innovativeness
D3	Data sets	Providing data with common formats such as tables or databases with quantifiable fields	High price elasticity, high importance of users' attribution
D4	Data processing service	Customized service on pre-processing such as data removal, and ETL (extraction- transformation- loading)	High price elasticity, medium innovativeness
D5	Data analysis tool services	Various algorithms and delivery methods, such as packages of algorithms, hardware, and software all-in-one	High price elasticity, high innovativeness
D6	Data application service	Customized industry solutions or portrait solutions for users, include data sources, service objects, and verification methods	High price elasticity, high innovativeness
D7	Combination of data goods/ services	Combination of data goods or services	Various, high importance of users' attribution

goods functions can meet users' needs by providing features that support their demands and stratify their specifications at the minimum threshold. These seven data goods are compared against each other, and their complexity can be categorized at two levels: high and low (with thresholds determined by domain experts). For example, the functions of the permit service and datasets are of low complexity, while API and data processing services are of high complexity. Empirical evidence suggests that product complexity negatively correlates with cost, time, and quality (Trattner et al., 2019).

Thus, the operational performance in terms of cost of data acquisition (X_1), time of data processing (X_2), and data quality (X_3) is used as the metric of this dimension, expressed as $F = f(X1, X2, X3)$, where $f(.)$ is a function, such as linear regressions or machine learning models with training data provided. The resulting value can end up with a closed range (e.g., [0,1]).

6.4   Case Study of Personalized Service on Telecom Data

This section presents a case study utilizing data samples from a DET service center in Southwest China. The statistics highlight the proportion of procurement methods and data sources for traded data. An instance of Telecom data demonstrates its delivery method through the item search API (for item-by-item queries).

Additionally, the characteristics of Telecom-data goods are analyzed and contrasted with those of e-book markets.

6.4.1   *Data Sample from a DET Service Center*

The case data were gathered from a DET service center in Southwest China on October 11, 2022. The dataset, with a sample size of 145, is a subset of their collection from December 17, 2021, to October 11, 2022. The statistics indicate that API is the traded data with the highest transaction volume (approximately 50.3%) in 2022, followed by data application services (e.g., business reports, 24.1%), data processing and analysis tool services (16.2%), and datasets (9.4%).

The statistics in Fig. 6.5a illustrate that the current mechanism for data trading involves bargaining in markets with reference price listings. The prices of data trading fluctuate based on data usage and quality. Among the five methods of data procurement, open tendering is the most prevalent, accounting for 67% of the procurement item number and 71.7% of the amount. Single-source procurement follows with 16.6% for items and 17.2% for amount, while competitive negotiation takes 11% for items and 4.9% for amount. Quotations inquiries represent 4.1% of the item number and 3.6% of the amount. The method of invited bidding is the least common, comprising 1.3% of the item number and 2.6% of the amount. Moreover, finance is the primary application field for traded data. In terms of procurement item number, banks procure the most (approximately 69.2%), followed by insurance (14.7%), securities (7.7%), and other sectors (8.4%).

The traded data (product) in the finance application domain encompasses various types of information, including personal and enterprise data. These data are predominantly (nearly 90%) sourced from public domains. For personal information data products, the primary components are personal credit/Telecom data, police records, and real estate data (see Fig. 6.5b). Enterprise information data products are more diverse, encompassing basic enterprise data, business activities, investment and financing, corporate profiles, affiliated enterprises, industry-specific information, and industrial chain data.

Many current data transactions eschew the use of data encryption algorithms that demand a high computational cost. High-speed computing enables rapid information sharing and decision-making during DET. The cost of a cluster node of computing power is approximately 200,000 Yuan per year. If the exchange service fee is set at 2% of the data trading volume, then the trading volume must exceed 20 million Yuan per year to cover the computational power cost. When additional costs such as transaction and operational expenses are factored in, the trading volume required is significantly higher.

The delivery methods of data products differ across the various classes of data goods. Telecom data is a critical source for digital financial applications (ITU, 2021). The subsequent sections describe the delivery of an item search API on

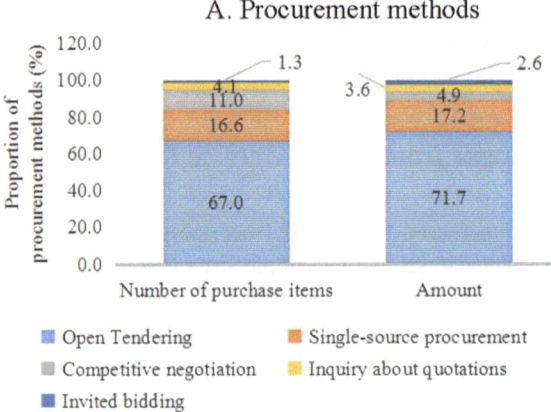

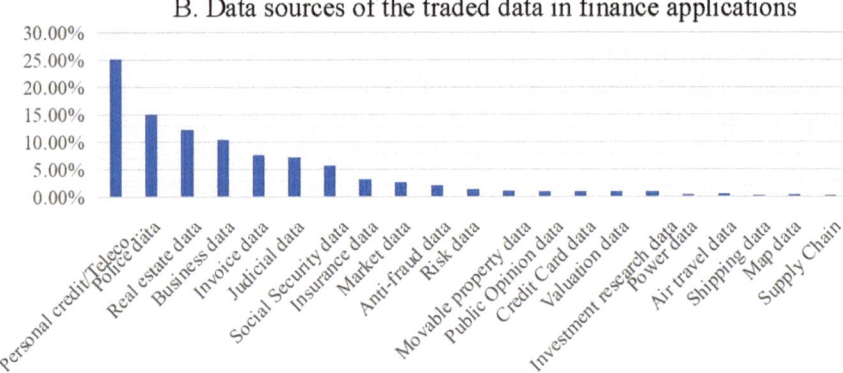

Fig. 6.5 Proportion of procurement methods and traded data sources (Data source: DET service center in 2022, Southwest China). (**a**) Procurement methods. (**b**) Data sources of the traded data in finance applications

Telecom data, providing empirical insights into the characteristics of Telecom-data goods and their market mode.

6.4.2 Telecom-Data Goods with Complexity and Specification

Telecom data for item-by-item queries exemplify the delivery method of the item search API. This involves three entities, telecom data suppliers, numerous data demanders, and exchanges, leading to a market-maker mode. The sample data of telecom-data goods are sourced from three Internet service providers (ISPs). Telecom-data goods primarily encompass two classes: permit service (D1) and

Table 6.4 Telecom-data goods with the combination of features

Data goods	Phone #	F1	F2	F3	F4	F5	F6	F7	Num.	Class of data goods
P01	√								1	D1
P02	√	√							2	D1
P03	√	√	√						3	D1
P11	√			√					2	D3
P12	√				√				2	D3
P21	√					√			2	D3
P22	√						√		2	D3
P23	√							√	2	D3

Note: F1–F7: name, ID #, length of registration time, average consumption in the past 3 months, ISP information of network transfer, status of use, whether secondary allocation. Num.: number of features

datasets (D3). The following section evaluates the characteristics of typical data goods using the case of Telecom-data goods.

The functions of Telecom-data goods often include various features. The complexity of these data goods varies with the number of their features. The combination of features involves selecting all or part of the features in a group, with the order of selection being irrelevant. These data goods can be generated by combining the features of block-wise data (see Table 6.4). In the case study's dataset, eight features (phone number and seven other features) can be combined into numerous data products. These eight products are described as $\{P01, P02, P03, P11, P12, P21, P22, P23\}$. The cost of data acquisition (X_1), time of data processing (X_2), and data quality (X_3) are set as constants for each unit of the Telecom data. The complexity of D1 is constant by using $F = f(X_1, X_2, X_3)$.

The time complexity of $D3$ is characterized by $n*m$ for data processing, where n is the sample size and m is the number of features. When the number of features is small, their functions are of low complexity; otherwise, their functions are of high complexity.

Telecom-data goods exhibit a low specification of demands for Telecom-permit service (D1) and either a high or low specification of demands for Telecom-data sets (D3). The specification of demands is determined by four factors: price (P), innovativeness (T), price elasticity (σ), and the importance (ω) of the Telecom-data goods. For Telecom-data goods with timely updating (see Table 6.5), their unit price ranges from 0.018 to 1.2 CNY/item (see Fig. 6.6a). The rate of updating varies among data goods. For instance, Telecom-data goods collected from three suppliers vary in update frequency from timely to monthly. Even among data goods from the same supplier, the update frequency varies. Data goods P12, P21, and P23 are updated in a timely manner, while P22 is updated T + 3 days. Data goods with the same features update at different frequencies when sourced from different suppliers, for example, the product P02 from China Unicom is updated T + 3 days, whereas the product from China Telecom is updated on the 15th of every month.

There are two contrasting scenarios for the influence of innovativeness (Panico & Cennamo, 2022). If consumers perceive the data goods as highly innovative, the

Table 6.5 Updating frequency, number of features, and price of data goods

Data goods supplier	Timely	T + 3 days	On 15th of every month	Monthly
Categories of Telecom-data goods				
China Mobile	P12, P21, P23	P22		P11, P02, P03
China Unicom	P12, P21, P23	P02, P03, P22		P11, P02, P03
China Telecom	P12, P21, P23	P22	P02, P03	P01, P11
Number of features, the importance (ω) of the Telecom-data goods				
China Mobile	2	2		2, 3
China Unicom	2	2, 3		2, 3
China Telecom	2	2	2	1, 2
Price (CNY/item)				
(lowest, mean, highest)	0.004, 0.23, 1.2	0.09, 0.13, 0.2	0.18, 0.20, 0.21	0.12, 0.23, 0.29

Note: T + 3 (days) indicates the delay within 3 days of the customers' transaction at ISP

parameter is set to 0.975; if innovativeness is undervalued, the parameter is set at $T = 0.525$. In the case of Telecom-data goods, the innovativeness is fixed at 0.525. The mean price elasticity (σ) is assumed to be -0.195 for mobile telephone services (Garbacz & Thompson Jr., 2007). The importance (ω) of Telecom-data goods is determined by the number of features. The results in Fig. 6.6 demonstrate how the price and importance (ω) of Telecom- data goods affect the specification of demands.

The results in Fig. 6.6c indicate that with the given importance (ω) of the Telecom-data goods, the frequency of updates has a direct impact on the specification of demands. A lower update frequency leads to a higher specification of demands, except for goods that are updated in real time. The price P and innovativeness T both have a positive effect on the specification of demands; the specification increases as the price rises.

Figure 6.6d, e reveals that the importance (ω) of the Telecom-data goods has a negative effect on the specification of demands when the base of the exponential function is less than 1, $T \cdot P^{-\sigma} < 1$. Conversely, the importance (ω) has a positive effect when the base of the exponential function is greater than 1. When the price P is approximately 0.727, the base of the exponential function reaches 1. However, the price data is anonymized during the data acquisition process by the data supplier.

6.4.3   Case of Market-Maker Mode

In this instance, we utilize the case of Telecom-data goods to assess the characteristics of one data market mode, the market-maker. When purchasing data goods, the process initially queries whether the Internet Service Provider (ISP) services the phone number (see Fig. 6.7a). If the answer is negative, the transaction is aborted; otherwise, the transaction proceeds. Once the query is resolved, the results of the

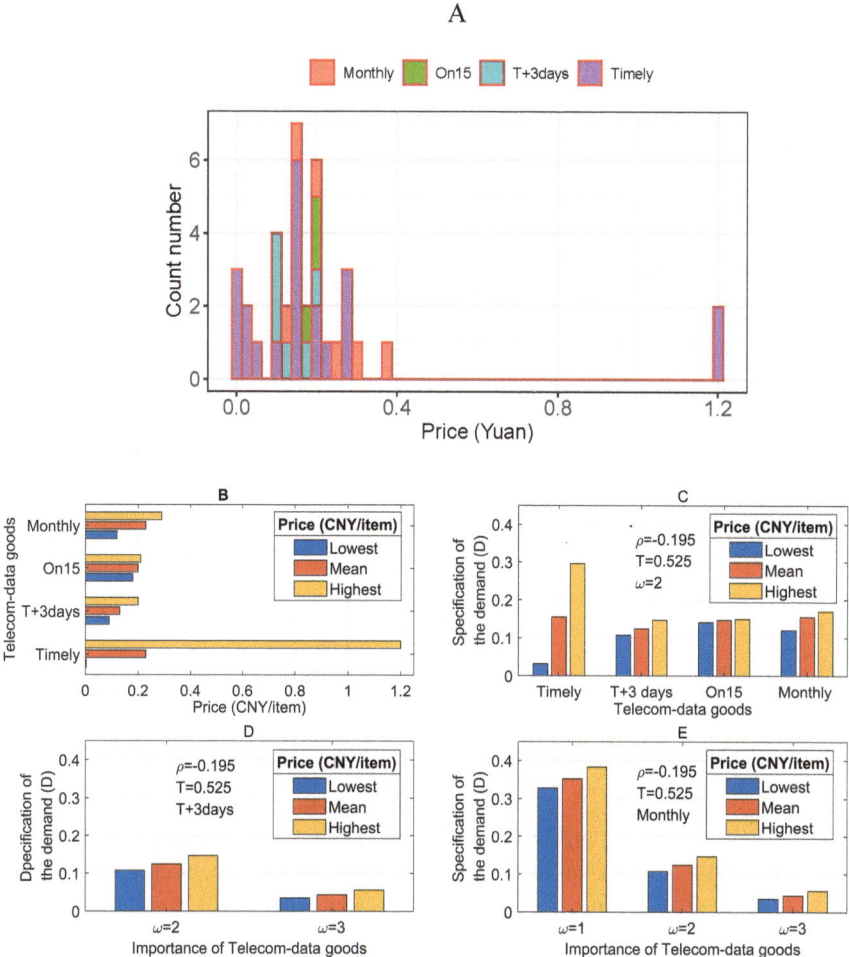

Fig. 6.6 Price distribution and demand specification of data goods. (**a**) Price distribution of 37 transactions from Dec 17, 2021, to October 11, 2022; (**b**) The lowest, mean, and highest price of the Telecom-data goods with four types of updating frequencies; (**c**) specification of demands vs. four types of updating frequencies; (**d**) specification of demands vs. importance of the Telecom-data goods for T + 3 days data goods; (**e**) specification of demands vs. importance of the Telecom-data goods for monthly data goods

data goods transaction are returned to the buyer, generating the billing. This DET mode thus possesses a strength in transparency, which facilitates supervision and is conducive to regulation. A significant proportion of current Telecom-data goods exhibit high security.

Diverse preferences are a common feature of the DET mode for Telecom-data goods. With respect to user preferences, the service first ascertains the task class. When the task necessitates verification, it determines the category of matching

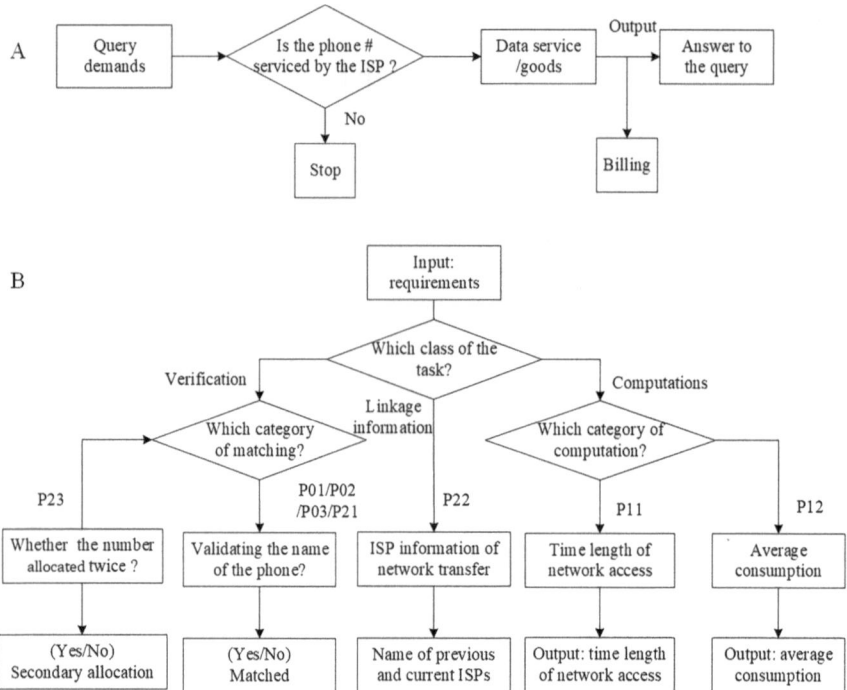

Fig. 6.7 Illustrative diagram of demand-driven data goods. (**a**) Telecom-data goods are associated with the service process. (**b**) Process of the Telecom-data service

required (see Fig. 6.7b). For verifying the phone number's name, data goods P01, P02, P03, and P21 are developed, and the outputs are the matching results: Yes or No. Similarly, for checking if a number is allocated twice (P23), similar query outputs are obtained.

For linkage information (P22), the task can retrieve the users' ISP information for network transfers. The outputs are the names of the previous and current ISPs. For computation goods (P11 and P12), the tasks can query the users' duration of network access and average consumption. Consequently, user preferences dynamically shift due to changes in the user base.

The current DET mode demonstrates a transition from information asymmetry to symmetry. The characteristics arise from the power imbalance in a transaction when telecom companies possess more information about the data goods than the other parties. For instance, a telecom company may have more information about the data quality, while downstream enterprises have more knowledge about how to utilize the data. This information asymmetry can lead to data buyers bidding insufficiently or telecom companies not knowing the appropriate value for data goods. In a mature market-maker mode, DET should demonstrate information symmetry in the contract design of data trading.

6.4.4   Comparing Data Goods with e-Book Markets

The characteristics of Telecom-data goods exchange are contrasted with electronic transaction markets (see Table 6.6). The current data marketplaces are relatively limited in scope compared to conventional e-business markets for physical goods. The nature of data goods renders their comparisons with electronic transaction markets distinct across several dimensions. We take e-books in electronic transaction markets as an example for comparison.

Regarding exchange models, Telecom-data markets operating under the market-maker mode can offer data goods of either high or low complexity. In contrast, the content of e-books is relatively uniform and can be considered a class of data goods with low complexity. In terms of the specification of demands, data markets provide services by dividing ownership (through data access) and use rights, with either high or low specificity of demands. Conversely, buyers of e-books often acquire product ownership and are typically licensed by book providers to others, often with low specificity of demands.

User preferences for Telecom-data goods frequently change dynamically. Data markets can facilitate DET services by aggregating Telecom-data suppliers and demanders. E-book transactions are typically served by an e-book seller or fragmented market entities. Information sharing in data markets is marked by data privacy and security concerns due to the digital nature of products (difficulty in confirming copyrights) and the convergence of value realization. For e-books, security concerns are relatively low, except in cases of copyright disputes. Furthermore, platforms for e-books generally possess more symmetrical information than data marketplaces.

Table 6.6   Comparing between Telecom-data goods and e-book sellers

Platform	Complexity of goods	Specification of demands	User preferences	Data security	Information sharing
Data market with Telecom-data goods	High or low complexity of data goods Many classes including datasets and permit services	High/low specificity of demands Service through division of ownership (different from traditional ownership), management and use rights	DET models aggregating data suppliers and demands, providing data transaction services, existing in all five models	Data privacy and security issues are prominent	Data products are characterized by challenging copyright authentication and value realization
e-book sellers	Low complexity Contents of e-books are relatively limited, viewed as one class of data goods	Low specificity of demands. Purchase of product ownership	e-book transactions for multiple buyers, not including M1 and M3, commonly served by fragmented market entities	Low-security concerns	Copyright disputation is important

Therefore, transactions on e-book platforms are currently more efficient. They provide insights into guiding participants in data markets, while further considerations must be given to designing an appropriate mechanism for enhancing the efficiency of data markets.

6.5 Discussion

The categorization of data market modes and data goods can assist DET managers in the development and management of new data goods and the collection of statistical data for market analysis. Data goods would be exchanged and traded efficiently through a single or unified data marketplace, catering to the diverse user preferences for data goods. Diverse preferences characterize the DET mode of Telecom-data goods. For instance, telecom companies are incentivized to provide access to their customer data for applications such as search, e-commerce, and social network profiling. Chinese policymakers have introduced numerous incentive plans at both national and provincial levels to enhance the efficiency of data exchange and trading. The 14th 5-Year Plan for National Informatization aims to establish systems for highly efficient data factor resource allocation, with the goal of enhancing data as a factor of endowment, activating the value of data, and creating new demand driven by data.

However, the limitations of the current market-maker mode are that it can be costly or inefficient during the initial stages of market development. As data exchanges are responsible for platform operations, they must adhere to data market governance principles, such as the Data Security Law. The heightened compliance obligations of data goods suppliers can limit their processing activities. Consequently, organizational and technical solutions should be established to empower data subjects to exercise their rights (e.g., the right of access and DET), enhancing the symmetry of information in data markets. When information sharing is asymmetrical, hidden behaviors or information often lead to market inefficiencies.

Most current data goods require high security. This prerequisite may dampen the ambition to produce rich Telecom-data goods, potentially leading to data monopolies. Nevertheless, an efficient DET can challenge data monopolies or even industry and regional monopolies by leveraging data as a factor of production. Future research will focus on other pivotal aspects for enhancing the efficiency of data markets, including the impact of data property rights on transaction costs.

References

Abbas, A. E., Agahari, W., van de Ven, M., Zuiderwijk, A., & de Reuver, M. (2021). Business data sharing through data marketplaces: a systematic literature review. *Journal of Theoretical and Applied Electronic Commerce Research, 16*(7), 3321–3339.

Agarwal, A., Dahleh, M., & Sarkar, T. (2019). A marketplace for data: An algorithmic solution. *Proceedings of the 2019 ACM conference on economics and computation.*

Ascarza, E., Ross, M., & Hardie, B. G. (2021). Why you aren't getting more from your marketing AI. *Harvard Business Review, 99*(4), 48–54.

Azcoitia, S. A., & Laoutaris, N. (2022). A survey of data marketplaces and their business models. *arXiv preprint arXiv:2201.04561*.

Azcoitia, S. A., Iordanu, C., & Laoutaris, N. (2021). What is the price of data? A measurement study of commercial data marketplaces. *arXiv preprint arXiv:2111.04427*.

Ba, S., Stallaert, J., & Whinston, A. B. (2001). Optimal investment in knowledge within a firm using a market mechanism. *Management Science, 47*(9), 1203–1219.

CAICT. (2023). *White Paper on the Data Factor (2022)*. China Academy of Information and Communications Technology.

Cappiello, C., Gal, A., Jarke, M., & Rehof, J. (2020). Data ecosystems: Sovereign data exchange among organizations (Dagstuhl Seminar 19391). *Dagstuhl Reports, 9*(9), 66–134.

Carvalho, A. (2020). A permissioned blockchain-based implementation of LMSR prediction markets. *Decision Support Systems, 130*, 113228.

Demchenko, Y., Los, W., & de Laat, C. (2018). Data as economic goods: Definitions, properties, challenges, enabling technologies for future data markets. *ITU Journal: ICT Discoveries, 2*(23), 1–10.

Deshmukh, H., Sundarmurthy, B., & Patel, J. M. (2020). To pipeline or not to pipeline, that is the question. *arXiv preprint arXiv:2002.00866*.

Garbacz, C., & Thompson, H. G., Jr. (2007). Demand for telecommunication services in developing countries. *Telecommunications Policy, 31*(5), 276–289.

Haupt, A., & Mugunthan, V. (2021). Prior-independent auctions for the demand side of federated learning. *arXiv preprint arXiv:2103.14375*.

Heidorn, H., & Weche, J. P. (2021). Business concentration data for Germany. *Jahrbücher für Nationalökonomie und Statistik, 241*(5-6), 801–811.

Huang, L., Dou, Y., Guo, M., Tang, Q., & Li, G. (2022a). Features and transaction modes of data products in data markets (in Chinese). *Big Data Research, 8*(3), 3–14.

Huang, S., Wan, X., Qiu, H., Li, L., & Yu, H. (2022b). Constrained optimization for stratified treatment rules with multiple responses of survival data. *Information Sciences, 596*, 343–361.

ITU. (2021). *Use of telecommunications data for digital financial inclusion*. I. T. Union.

Jiao, Y., Wang, P., Niyato, D., Alsheikh, M. A., & Feng, S. (2017). Profit maximization auction and data management in big data markets. *2017 IEEE wireless communications and networking conference (WCNC)*.

Jones, C. I., & Tonetti, C. (2020). Nonrivalry and the economics of data. *American Economic Review, 110*(9), 2819–2858.

Ke, T. T., & Sudhir, K. (2023). Privacy rights and data security: GDPR and personal data markets. *Management Science, 69*(8), 4389–4412.

Kingaby, S. A. (2022). *The stock market API*. Springer.

Kolaitis, P. G. (2018). Reflections on schema mappings, data exchange, and metadata management. *Proceedings of the 37th ACM SIGMOD-SIGACT-SIGAI symposium on principles of database systems*.

Kozyreva, A., Lorenz-Spreen, P., Hertwig, R., Lewandowsky, S., & Herzog, S. M. (2021). Public attitudes towards algorithmic personalization and use of personal data online: Evidence from Germany, Great Britain, and the United States. *Humanities and Social Sciences Communications, 8*(1), 1–11.

Lauga, D. O., & Ofek, E. (2011). Product positioning in a two-dimensional vertical differentiation model: The role of quality costs. *Marketing Science, 30*(5), 903–923.

Li, Y., Ping, Y., Zhong, Y., & Misra, R. (2023). Learning-by-doing in non-homogeneous tasks: An empirical study of content creator performance on a music streaming platform. *Electronic Commerce Research and Applications, 58*, 101241.

Liu, Y.-L., Yan, W., & Hu, B. (2021). Resistance to facial recognition payment in China: The influence of privacy-related factors. *Telecommunications Policy, 45*(5), 102155.

Panico, C., & Cennamo, C. (2022). User preferences and strategic interactions in platform ecosystems. *Strategic Management Journal, 43*(3), 507–529.

Pei, J. (2020). *A survey on data pricing: From economics to data science.*

Qin, X., Liu, Z., & Tian, L. (2021). The optimal combination between selling mode and logistics service strategy in an e-commerce market. *European Journal of Operational Research, 289*(2), 639–651.

Rasouli, M., & Jordan, M. I. (2021). Data sharing markets. *arXiv preprint arXiv:2107.08630.*

Rodríguez-Crespo, E., & Martínez-Zarzoso, I. (2019). The effect of ICT on trade: Does product complexity matter? *Telematics and Informatics, 41*, 182–196.

Sadowski, J. (2019). When data is capital: Datafication, accumulation, and extraction. *Big Data and Society, 6*(1), 1–12.

Shen, M., Tang, C. S., Wu, D., Yuan, R., & Zhou, W. (2020). JD.com: Transaction-level data for the 2020 MSOM data driven research challenge. *Manufacturing and Service Operations Management.* https://doi.org/10.1287/msom.2020.0900

Slater, S. F., & Narver, J. C. (1999). Market-oriented is more than being customer-led. *Strategic Management Journal, 20*(12), 1165–1168.

Spiekermann, M. (2019). Data marketplaces: Trends and monetisation of data goods. *Intereconomics, 54*(4), 208–216.

Taylor, L., Mukiri-Smith, H., Petročnik, T., Savolainen, L., & Martin, A. (2022). (Re) making data markets: An exploration of the regulatory challenges. *Law, Innovation and Technology*, 1–40.

Thouvenin, F., Tamò-Larrieux, A., & Burri, M. (2021). Data ownership and data access rights: Meaningful tools for promoting the European digital single market. In M. Burri (Ed.), *Big data and global trade law* (pp. 316–339). Cambridge University Press.

Tian, L., Vakharia, A. J., Tan, Y., & Xu, Y. (2018). Marketplace, reseller, or hybrid: Strategic analysis of an emerging e-commerce model. *Production and Operations Management, 27*(8), 1595–1610.

Trattner, A., Hvam, L., Forza, C., & Herbert-Hansen, Z. N. L. (2019). Product complexity and operational performance: A systematic literature review. *CIRP Journal of Manufacturing Science and Technology, 25*, 69–83.

Treleaven, P., Barnett, J., Knight, A., & Serrano, W. (2021). Real estate data marketplace. *AI and Ethics, 1*(4), 445–462.

Udry, C. (1996). *Efficiency and market structure: Testing for profit maximization in African agriculture.* Department of Economics, Northwestern University.

Wattal, S., Telang, R., & Mukhopadhyay, T. (2009). Information personalization in a two-dimensional product differentiation model. *Journal of Management Information Systems, 26*(2), 69–95.

Yu, H., & Chen, J. (2022). Treatment effect identification using two-level designs with partially ignorable missing data. *Information Sciences, 611*, 277–300.

Yu, H., Yang, C.-C., & Yu, P. (2023a). Constrained optimization for stratified treatment rules in reducing hospital readmission rates of diabetic patients. *European Journal of Operational Research, 308*(3), 1355–1364. https://doi.org/10.1016/j.ejor.2022.12.020

Yu, H., Zuo, X., Tang, J., & Fu, Y. (2023b). Identifying causal effects of the clinical sentiment of patients' nursing notes on anticipated fall risk stratification. *Information Processing and Management, 60*(6), 103481.

Zhu, B., Bates, S., Yang, Z., Wang, Y., Jiao, J., & Jordan, M. I. (2022). The sample complexity of online contract design. *arXiv preprint arXiv:2211.05732.*

Chapter 7
Ghost Data in Data Quality Management

7.1 Ghost Data in Statistics

Within the domain of quantum field theory, a ghost (Davies & Brown, 1993), alternately termed a ghost field or a gauge ghost (Scharf, 2001), represents a non-physical state in gauge theories. These states are crucial for preserving gauge invariance in situations where the number of local fields transcends the actual physical degrees of freedom.

The classification "ghost data" refers to a spectrum of data types, such as missing data, data generated through simulations, pretend data, virtual data, and highly sparse data, as detailed in Table 7.1.

Missing data is a concept familiar to researchers, and its handling requires diverse methods due to the various mechanisms underlying data absence (Little & Rubin, 2019), including missing at random (MAR) and missing not at random (MNAR). Techniques for addressing missing data predominantly encompass sampling inference, Bayesian inference, and likelihood-based methods. In the context of experimental design, random complete block (RCB) designs may be susceptible to missing data (George et al., 2005). A (partially) balanced incomplete block design can be conceptualized as an RCB design with missing data for the purpose of statistical analysis.

Figure 7.1 presents illustrative examples of ghost data. Sherlock Holmes is renowned for his ability to discern truth by meticulously analyzing evidence, such as charts or evidence chains. For instance, the absence of a dog's bark upon encountering a suspect can suggest to Holmes that the dog is familiar with the individual, thus narrowing the pool of suspects. In the realm of global health research, evidence gaps and the proliferation of counterfeit drugs (Hodges & Garnett, 2020) have been identified as significant issues. Consequently, the international health research and policy communities have sounded the alarm regarding the escalating danger posed by counterfeit and low-quality medications, collectively referred to as fakes.

Table 7.1 Ghost data and examples in statistics

Subsets	Characteristics	Examples
Missing data	Counterfactual outcomes or missing data (especially non-ignorable) in causal inference (Rubin, 2005; Yu et al., 2020), sometimes highly sparse	Holmes' evidence chain (Wikipedia, 2020a–09-19) aircraft dropouts in battles (Mangel & Samaniego, 1984)
Simulation data	Model simulation to provide solutions for "what if"; simulation in manufacturing systems or survival as a digital ghost (Steinhart, 2007)	Flight simulator or virtual patients with heart disease (Xu et al., 2017)
Pretend data	Fake news; scientific misconduct; falsified data	Low-quality drugs (Hodges & Garnett, 2020) Jiang Gan Steals a Letter (an episode from the *Three Kingdoms*) (Luo, 2018)
Virtual data	Digital information created by the activity of robots, intelligent agents, computers, mobile phones, embedded systems, and other networked devices	AlphaZaro (Silver et al., 2018) Generated faces (while not often seen in humans) from decoding with Generative Adversarial Networks (Lei et al., 2020)
Highly sparse data	The data contains a lot of empty or unused space	Text mining, recommendation systems, genomics, and social networks

(a) Holmes' evidence chain
(wikipedia, 2020-09-19a)

(b) I See Dead People
(wikipedia, 2020-09-19b)

Fig. 7.1 Examples of ghost data from the statistics perspective. (**a**) Holmes' evidence chain (Wikipedia, 2020a–09-19). (**b**) "I See Dead People" (Wikipedia, 2020b–09-19)

Additionally, ghost data encompasses generative machine data, which is produced by the activities of robots, intelligent agents, computers, mobile phones, embedded systems, and other networked devices. For instance, the decoding process utilizing generative adversarial networks (Lei et al., 2020) frequently yields objects or faces that are not commonly encountered in the real world or recognized by humans. Another relevant example is the narrative "I See Dead People" from *The Sixth Sense* (Wikipedia, 2020b–09-19), which illustrates how individuals may describe perceptions of phenomena that are not visibly apparent to others.

Highly sparse data refers to a dataset with a significant portion of zero elements or null values. In other words, the data contains a lot of empty or unused space. For example, in text mining and large language models, when text data is converted into numerical format (e.g., using term frequency-inverse document frequency), the output matrix is often sparse because most words do not appear in many documents. For another example, in collaborative filtering, the user-item interaction matrix is typically sparse because users only interact with a small subset of available items.

7.2   Ghost Data in Computer Science

Ghost data encompasses various types of data, akin to metaphorical dark matter and dark energy within the realm of computer vision, as well as digital art, digital museum content, ghost imaging techniques, and instances of data hallucinations, as detailed in Table 7.2.

Table 7.2 Ghost data and examples in computer science

Subsets	Characteristics	Examples
Dark matter/ energy in computer visions	Entities such as objects, stuff like liquid, human actions, scenes, and relations, such as intents of humans, physical fields, and attractions in a scene, cannot be recognized by the geometry and appearance features commonly used in current computer vision research	Dark matter in video scenes (Xie et al., 2017)
Digital art	Virtual reality, augmented reality, video production in scientific movies or esports games	Movie *Edge of Tomorrow*; emotion (cognition) in the text, images, and videos sensible for the right person
Digital museum	Digital craftsmanship that restores history, a digital entity that draws on a museum's characteristics to complement, enhance, or augment the museum experience through personalization, interactivity, and richness of content	Virtualization of the terracotta warriors (CCTV, 2020-09-19)
Ghost imaging	Producing an image of an object by combining information from two light detectors: a conventional, multi-pixel detector that doesn't view the object, and a single-pixel (bucket) detector that does view the object	Visualization of ghost imaging (Khakimov et al., 2016)
Hallucinations	The model generates text that is inconsistent with the input context or does not correspond to any real-world knowledge	Model produces data that are factually incorrect, logically incoherent, or simply not grounded in reality

Note: CCTV refers to China Central Television

In the computer vision context (Xie et al., 2017), the metaphorical "dark matter" within video scenes (Fig. 7.2a) is differentiated from other objects primarily by its functional ability to attract or repel people, rather than by its visual appearance. Within digital museums, visitors can engage with digital artisanship that reconstructs historical artifacts or observe the processes by which entities were created or operated in the past. In the cultural domain of virtual restoration, data scientists are able to encode additional information and retrieve latent details in collaboration with heritage conservation experts. For instance, virtual reality techniques (Fig. 7.2b) have been employed to virtually replicate the terracotta warriors (CCTV, 2020-09-19), offering a near-perfect facsimile of their original state. As technology advances, it allows not only for the preservation of history but also for the persistence of digital ghosts (Steinhart, 2007)—entities that live on in diaries, photographs, audio recordings, and films. Effectively, one can "live on" after death through various digital artifacts or simulators, such as virtual patients experiencing heart disease-related pain (Fig. 7.2c). Furthermore, quantum ghost imaging

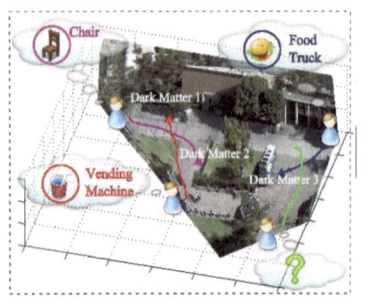

(a) "Dark Matter" in video scenes (Xie et al., 2017)

(b)Virtualization of the terracotta warriors (CCTV, 2020-09-19)

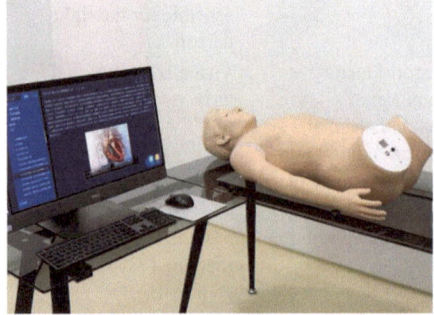

(c) Virtual patients with Heart disease (Xu et al., 2017)

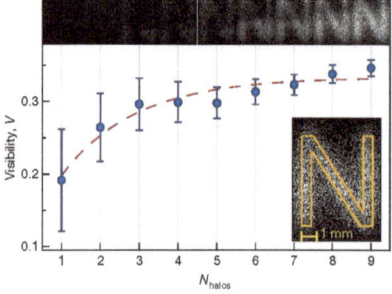

(d)Visualization of ghost images (Khakimov et al., 2016)

Fig. 7.2 Examples of ghost data from the computer science perspective. (**a**) "Dark matter" in video scenes (Xie et al., 2017), (**b**) virtualization of the terracotta warriors (CCTV, 2020-09-19), (**c**) virtual patients with heart disease (Xu et al., 2017), (**d**) visualization of ghost images (Khakimov et al., 2016)

technologies (Khakimov et al., 2016) are capable of generating images of an object by merging data from two light detectors: a conventional, multi-pixel detector that does not directly observe the object and a single-pixel (bucket) detector that does (Fig. 7.2d).

In the domain of physics, the term "ghosts" refers to conceptual entities that are categorized into "good ghosts," such as Faddeev-Popov ghosts, and "bad ghosts," such as Pauli-Villars ghosts. This classification is mirrored in the concept of ghost data, where there are analogous entities that can be considered "bad ghosts." Additionally, certain types of ghost data may encode emotional or cognitive content within text, images, and videos, making them discernible only to the intended audience. Although such records may capture only a subset of superficial features, certain individuals may interpret them as containing more profound and pertinent information, while others may not. Individuals capable of detecting the embedded layers of meaning can potentially reconstruct elements of their life narratives and fashion approximate replicas of their personal experiences.

7.3   Ghost Data in Economics

The term "invisible hand" refers to the unseen mechanisms that guide the free market economy. Coined by Adam Smith in 1759, it encapsulates the idea that self-interested individuals, acting within a system of mutual reliance in a free market economy, inadvertently promote the overall welfare of society. Through the pursuit of their own self-interests and the freedom to produce and consume as they choose, the collective interests of society are ostensibly served. This interdependence motivates producers to supply goods and services that are socially needed, despite their personal concerns. The concept suggests that markets can reach equilibrium naturally, without the need for government intervention or other external influences that might distort natural market dynamics. Each voluntary exchange communicates information about the value and scarcity of goods and services. However, critics contend that the invisible hand does not always result in outcomes that are beneficial to society and can sometimes foster greed, negative externalities, inequality, and other detrimental effects.

The concept of hidden actions, also known as moral hazard, posits that contracts should remunerate agents for their efforts to achieve favorable outcomes on behalf of the principal. Consequently, these contracts must be structured to rely on observable and verifiable measures of outcome quality, such as the actual accuracy of a classification model. This poses a significant challenge in machine learning (ML) algorithms, where the accuracy of the learned model is not known in advance and can be subject to randomness. The principal may choose to invest in resources, such as extensive test datasets, to mitigate this uncertainty and more effectively assess the agent's contributions. A key consideration in this context is whether the principal should focus on precisely verifying outcomes or instead design incentives that motivate the agent to ensure high-quality results from the outset (Table 7.3).

Table 7.3 Ghost data and examples in economics

Subsets	Characteristics	Examples
Invisible hands	The unseen mechanisms that guide the free-market economy	Self-interested individuals acting within a system of mutual reliance in a free market economy, inadvertently promote the overall welfare of society
Moral hazard	Determining whether the principal must accurately verify the outcome or instead incentivize the agent in the first place to ensure a high-quality outcome	The accuracy of the learned model is unknown a priori and random
Adverse selection	Whether contracts exist that appropriately incentivize the agent to perform his best while knowing very little about the optimal possible accuracy	In ML algorithms, the true error of the best model is unknown

The concept of hidden state, also known as adverse selection, suggests that optimal contracts leverage knowledge of the best achievable outcome to motivate agents to strive for those outcomes. This is particularly difficult in machine learning (ML) algorithms, where the true error rate of the best possible model is not known in advance. Moreover, methods that estimate the optimal accuracy often require resources comparable to those needed to develop a model with that level of accuracy. Thus, the challenge lies in designing contracts that effectively incentivize agents to deliver their best performance when there is minimal information available about the highest possible accuracy achievable.

7.4 Further Discussion

The procurement of high-quality data encompasses a multitude of stages, including data collection, annotation, purification, de-identification, aggregation, and analysis. It entails research into the standardization of data resources and the enhancement of quality for micro-level entities, in addition to facilitating data interchange, collaborative efforts, and open scientific inquiry among different entities.

Nowadays, decentralized data marketplace (Ananthakrishnan et al., 2023) or exchange presents a promising avenue for further exploration. These platforms are built upon secure, smart contract-enabled blockchains, which empower users to reclaim ownership of their data and leverage the economic value of their data attributes without relying on a centralized governance structure.

The following insights underscore three critical considerations:

1. First, researchers must clarify the provenance of the data.
2. Second, data collected for one purpose may not be suitably repurposed without careful evaluation.

3. Third, Deming's principles maintain their relevance amidst advancements in computation and automation. If Deming were present in the age of big data, one might ponder how he would confront the challenges it poses and what novel insights and strategies he might propose.

However, US policymakers tend to sometimes overlook valuable advice. For instance, Deming's ideas were initially embraced, adopted, and expanded by the Japanese, elevating him to legendary status, while the United States was slow to recognize their value. Similarly, Air Force Major John Boyd (Osinga, 2007) proposed the Lightweight Fighter program. Yet, it was the Navy that first took notice of his ideas, leading to his achieving legendary status within the Navy rather than in his home branch, the Air Force.

Statisticians are integral to the advancement of data science. While the data science community has made progress, there is still a widespread need for a comprehensive grasp of core statistical concepts. For instance, obtaining accurate results often necessitates an adequate sample size. Model building is an iterative endeavor, where the objectives of the model dictate the appropriate methodologies. Not every analytical challenge is predictive in nature. Establishing a solid theoretical foundation is critical for a sustained technological development process. Data encompasses not only the descriptive aspects of the situation but also the methodologies for its manipulation and analysis.

Data scientists frequently command expertise in areas that may not be as broadly recognized within the statistical community. The complexities of data storage, retrieval, and processing in the context of big data require a high level of technical understanding, which may not be as widely distributed among statisticians. The business sector has warmly embraced programming, acknowledging the pivotal role of algorithms in achieving organizational objectives. For example, the empirical utility of random forest methods has surpassed their theoretical foundations. Furthermore, specialized domain knowledge is indispensable in the nuanced practice of data science.

In summation, statistics is a pivotal discipline underpinning computer science, machine learning, and artificial intelligence (CS/ML/AI). Drawing an analogy from Darwin's theory of evolution, it is not necessarily the strongest species that prevails but rather the most adaptable. This concept is equally applicable to the progressive development and metamorphosis of data science and statistical methodologies.

References

Ananthakrishnan, N., Bates, S., Jordan, M. I., & Haghtalab, N. (2023). Delegating Data collection in decentralized machine learning. *arXiv preprint arXiv:2309.01837*.

CCTV. (2020-09-19). *Artificial intelligence and AR technology restores terracotta warriors*. http://shaoer.cctv.com/2017/06/29/ARTIohn1G4uBr8U27lMo7X4k170629.shtml

Davies, P. C. W., & Brown, J. R. (1993). *The ghost in the atom: A discussion of the mysteries of quantum physics*. Cambridge University Press.

George, E., Hunter, J. S., Hunter, W. G., Bins, R., Kirlin, K., IV, & Carroll, D. (2005). *Statistics for experimenters: Design, innovation, and discovery*. Wiley.

Hodges, S., & Garnett, E. (2020). The ghost in the data: Evidence gaps and the problem of fake drugs in global health research. *Global Public Health*, 1–16.

Khakimov, R. I., Henson, B., Shin, D., Hodgman, S., Dall, R., Baldwin, K., & Truscott, A. (2016). Ghost imaging with atoms. *Nature, 540*(7631), 100–103.

Lei, N., An, D., Guo, Y., Su, K., Liu, S., Luo, Z., Yau, S.-T., & Gu, X. (2020). A geometric understanding of deep learning. *Engineering*.

Little, R. J., & Rubin, D. B. (2019). *Statistical analysis with missing data* (Vol. 793). Wiley.

Luo, G. (2018). *The romance of the three kingdoms*. Penguin.

Mangel, M., & Samaniego, F. J. (1984). Abraham Wald's work on aircraft survivability. *Journal of the American Statistical Association, 79*(386), 259–267.

Osinga, F. P. (2007). *Science, strategy and war: The strategic theory of John Boyd*. Routledge.

Rubin, D. B. (2005). Causal inference using potential outcomes: Design, modeling, decisions. *Journal of the American Statistical Association, 100*(469), 322–331.

Scharf, G. (2001). *Quantum gauge theories: A true ghost story*.

Silver, D., Hubert, T., Schrittwieser, J., Antonoglou, I., Lai, M., Guez, A., Lanctot, M., Sifre, L., Kumaran, D., & Graepel, T. (2018). A general reinforcement learning algorithm that masters chess, shogi, and Go through self-play. *Science, 362*(6419), 1140–1144.

Steinhart, E. (2007). Survival as a digital ghost. *Minds and Machines, 17*, 261–271.

Wikipedia. (2020a-09-19). *Sherlock Holmes*. Retrieved 2020-09-19 from

Wikipedia. (2020b-09-19). *The Sixth Sense*. Retrieved 2020-09-19 from

Xie, D., Shu, T., Todorovic, S., & Zhu, S.-C. (2017). Learning and inferring "dark matter" and predicting human intents and trajectories in videos. *IEEE Transactions on Pattern Analysis and Machine Intelligence, 40*(7), 1639–1652.

Xu, M., Shen, J., & Yu, H. (2017). *Big data medical: Medical intelligence in the era of cognitive science*. Machinery Industry Press.

Yu, H., Wang, Y., Wang, J.-N., Chiu, Y.-L., Qiu, H., & Gao, M. (2020). Causal effect of honorary titles on physicians' service volumes in online health communities: Retrospective study. *Journal of Medical Internet Research, 22*(7), e18527.

Chapter 8
Summary

High-quality data has emerged as a critical factor of production, playing a pivotal role in fostering digital economic growth. The acquisition of such data is essential for contemporary decision-making frameworks.

The primary contributions of this book are as follows: Firstly, it systematically reviewed the methods, pillars, and tools of data quality management. This book provides an overview of the evolution and cutting-edge techniques in data quality control. Data quality forms the foundation of datasets, as machine learning algorithms inherently encounter uncertainties such as measurement errors. The two predominant guidelines for gathering high-quality data are randomization and systematicity. Regardless of the method employed—experimental or research design—these principles are central. Intelligent data acquisition methods are universally applicable and should be integrated into empirical research whenever feasible. Even conventional techniques can yield new insights from high-quality data. Experimental design and statistical quality control methodologies can inform the data selection process. Hence, maintaining data quality is essential to ensure that the service meets user expectations.

Secondly, this book extends conventional data quality management methods to new enriching scenarios, such as data science and data markets. These two scenarios provide rich contexts in which data quality problems occur and need to be prevented or controlled. The development of data science benefits from the new insights into quality management that statistics can provide. Data markets provide a mechanism to enhance data availability, particularly in data-scarce domains such as personalized medicine and services. Service personalization leverages user data to tailor offerings and improve outcomes. Personal data is used to predict needs and requirements for customization, with data processing and the integration of AI technologies playing crucial roles. Consequently, potential data providers are incentivized to enter the market. A significant challenge for data buyers in these markets is identifying high-quality and valuable data points among various sellers.

H. Yu, *Data Quality Management in the Data Age*, SpringerBriefs in Service Science, https://doi.org/10.1007/978-3-031-71871-7_8

Thirdly, it conducted a qualitative examination of the factor market for data exchange and trading. This study synthesized the modes of data exchange and trading with a factor design. The case study presented empirical results for the market-maker mode with Telecom-data goods. Data markets offer a mechanism for the procurement of high-quality data and for augmenting the overall data supply. Data market managers and regulatory bodies must establish an environment that incentivizes potential data sellers, particularly those with high-quality data, to participate in the market. Consequently, data scientists and engineers operating within these markets must enhance their proficiency in data quality management. Therefore, high-quality data can accelerate the growth of the digital economy.